Mourning in the Anthropocene

MOURNING IN THE ANTHROPOCENE

Ecological Grief and Earthly Coexistence

Joshua Trey Barnett

Michigan State University Press | *East Lansing*

♾ The paper used in this publication meets the minimum requirements
of ANSI/NISO Z39.48-1992 (R 1997) (Permanence of Paper).

Michigan State University Press
East Lansing, Michigan 48823-5245

Library of Congress Cataloging-in-Publication Data
Names: Barnett, Joshua, author.
Title: Mourning in the Anthropocene : ecological grief and earthly coexistence /
Joshua Trey Barnett.
Description: East Lansing, Michigan : Michigan State University Press, [2022] |
Includes bibliographical references and index.
Identifiers: LCCN 2021052814 | ISBN 978-1-61186-434-2 (Paperback : acid-free paper) |
ISBN 978-1-60917-704-1 (PDF) | ISBN 978-1-62895-472-2 (ePub) |
ISBN 978-1-62896-466-0 (Kindle)
Subjects: LCSH: Environmental degradation—Psychological aspects. |
Ecological disturbances—Psychological aspects. | Global environmental change—
Psychological aspects. | Grief. | Loss (Psychology) | Nature—Effect of human beings on.
Classification: LCC GE140 .B37 2022 | DDC 304.2/8—dc23/eng/20220110
LC record available at https://lccn.loc.gov/2021052814

Cover design by Anastasia Wraight
Cover art: Robert Havell Jr. engraving of *John J. Audubon, Passenger Pigeon*, 1829.
Gift of Mrs. Walter B. James, Courtesy National Gallery of Art, Washington.

Contents

Acknowledgments

Writing surely requires solitude. But it is not, for that reason, a lonely undertaking. Writing is, rather, an activity undertaken in connection and conversation with others—some living, some dead; some nearby, some distant; some of whom one knows, some of whom one has never met. And so, though it is my name that appears on the cover of this text, and though I assume responsibility for what follows, the ideas and arguments presented here developed out of the relations in which I have found myself entangled over the last few years. For the thinking and writing these relations have made possible, I am endlessly grateful.

I have been blessed with caring, thoughtful, and supportive teachers. Among others, Jamie Landau, Kelly Happe, and Michelle Ballif introduced me to the intellectual joys and challenges of rhetorical criticism while I studied at the University of Georgia. In Athens, Georgia, Corey Johnson played a vital role in my early intellectual development and cultivated my identity not only as a scholar but also as a writer. Without their guidance and encouragement, I would not be where—or *who*—I am today. While studying at Indiana University, Bob Ivie, Phaedra Pezzullo, Robert Terrill, Anjali Vats, and Susan Lepselter created an intellectual environment that was equal parts stimulating, rigorous, and exciting. At the University of Utah, I was fortunate to work closely with several remarkable teachers, including Helga Shugart, Robin Jensen, Danielle Endres, Steve Tatum, and Esther Raskin, each of whom made significant, enduring impressions on my scholarly identity and interests. It is to Kevin DeLuca, though, that I owe the most: as an inspiring teacher, wise mentor, and kind friend, Kevin showed me—and, indeed, continues to show me—how and why to lead a life of the mind.

I began the research which led to this book while living near the north shore of Lake Superior and teaching at the University of Minnesota, Duluth. There I had the privilege of working alongside several thinkers who helped me find my way to this project: David Gore, Michael Pfau, Randel Hansen, Kathryn Milun, and Aaron Boyson welcomed me into their Anthropocene study group early on and gave me a great deal to think about during my three years in Duluth. As dean of the College of Liberal Arts, Susan Naramore Maher provided encouragement and institutional support beyond what any assistant professor could expect, including a semester-long teaching release to read and write my way into this book project. That time made all the difference. While in Minnesota, I also had the good fortune to teach several courses in which the themes of this book were central. Students in those seminars—especially Oliver Cygan, Cameron Olson, and Addison Ferrie—listened to and provided invaluable feedback on the ideas as they were developing.

I wrote and revised most of this book during my first year at Pennsylvania State University. I joined the faculty of Communication Arts and Sciences in July of 2020; when I arrived, the COVID-19 pandemic was raging across the globe. While navigating the many obstacles imposed by the virus, my colleagues here in central Pennsylvania welcomed and supported and encouraged me. Collectively, they have made it possible for me to bring this project to fruition. Denise Solomon's vision, wisdom, stability, and kindness have been vital. I especially appreciate my rhetoric colleagues—all of whom have cheered me on and many of whom have listened to, helped, and mentored me as I worked on this project. Mary Stuckey, Pamela VanHaitsma, Rosa Eberly, Kirt Wilson, Jeremy Engels, Stephen Browne, Katya Haskins, Michael Steudeman, Michele Kennerly, Debbie Hawhee: thank you for the kind embrace and the intellectual stimulation. I especially want to thank Mary Stuckey, who read parts of this manuscript and encouraged me to lean into my own voice, and Pamela VanHaitsma, whose company and conversation while walking, hiking, and writing have been energizing. While revising this project, Cheyenne Zaremba provided research assistance. For their diligence and exuberance, I thank them. At Pennsylvania State University, I am also enormously grateful to the Huck Institutes of the Life Sciences for seeing the value in and supporting my research.

I feel extremely lucky to have worked with the Michigan State University Press on this project. Above all, I must thank my editor, Catherine Cocks, whose enthusiasm for this book has kept me going and whose experience and wisdom have made a daunting process seem doable at every step of the way. Two anonymous reviewers offered incisive feedback on the project; this is undoubtedly a better book because of their careful readings and astute advice. Wherever you are, whoever you are: thank you! For her work creating the index at the end of this book, I extend a hearty thanks to Dawn Martin.

In too many ways to elaborate here, colleagues near and far guided and supported me while I was writing this book. They asked smart questions, shared readings, read drafts, and otherwise encouraged me to carry on the work. For all that and more, thank you: Leland Spencer, Chuck Morris, Jamie Landau, Betsy Brunner, Chris Ingraham, Devika Chawla, John Lynch, Jen Clary-Lemon, Marek Muller, Julie Madrone Schutten, Emma Bloomfield, Rebecca de Souza, Greg Seigworth, Eric Jenkins, Jeremy Gordon, Kirstin Wagner, Carolyn Pedwell, Liz Parks, Catherine Bruns, Tyler Rife, and Christopher Jackson. I have undoubtedly omitted many other people who, either directly or indirectly, have made an impression on me while writing this book: thank you.

Outside the academy, several friends have sustained me while working on this book. For all the conversations, visits, care packages, and trips together over the years, thank you: Meg Graham, Jasmine Morrissette, Brandon Hill, Erick Amick, Yancey Gulley, Spencer Turner, and Clint Urvand.

How, finally, to thank my family—biological and chosen? To my parents, Tammy and Trey Barnett: thank you for teaching me the values of hard work and persistence, and for your unconditional love. To my sister, Malorie Hughes, and her beautiful family—Peter, James Douglas, Nora, and Haddie—thank you for your enthusiastic support. And to the one without whom I cannot imagine living, my Brandon Killen: thank you for your unwavering support, thank you for your companionship, thank you for your honesty, thank you for the good meals, thank you for putting up with me, thank you for your levity, and, most of all, thank you for your love—and for your care.

PROLOGUE

Loss

A Reckoning

Earthly coexistence entails loss. It has always been this way—and so it shall remain. Beings and ways of being come into and pass out of existence. Landscapes and ecosystems emerge and wither away. Ecological and earth systems develop and become otherwise. An inherent dimension of life on earth, such losses cannot be willed away.

Nevertheless, some of us are better able to forget, disregard, or deny ecological and earthly losses in our day-to-day lives. This capacity to overlook is the product of privileges about which we may be more or less aware, privileges which are bound up with class, race, nationality, culture, and much else besides. When forced to acknowledge ecological losses and earthly transformations, those of us who share in these privileges frequently treat them as though they were phenomena distinctive to the present and in urgent need of response—that is, as "crises" in need of expedient resolution.[1] Rather than sitting with such losses, we turn quickly to the work of fixing what appears to have been broken in hopes that we might soon be able to forget again the destruction, the transformation, the loss.

For others, however, the extinction of more-than-human species, the destruction of landscapes, and the disruption of land communities organize everyday life. Understood as both historical and ongoing, ecological and earthly losses press down especially forcefully on those among us whose way of dwelling is intertwined with specific plants and animals, with particular landscapes and ecosystems, and with recurrent weather patterns and annual cycles. After the United States government had all but terminated the buffalo (*Bison bison*) in an attempt to deprive Native Americans of their traditional lifeways in the nineteenth century, for instance, Chief Plenty-Coups (Crow) remarked that "when the buffalo went away the hearts of my people fell to the ground, and they could not lift them up again. After this nothing happened."[2] As buffalo populations waned, so too did the very conditions for what it meant to be an Indian on the Plains. Especially when one's sense of who one is, as well as one's capacity to sustain one's way of life, is bound up with what has been lost, overlooking the loss simply is not possible.

Of course, not all ecological and earthly losses are equal. Humans sometimes enact violence against more-than-human beings and ways of being, as the case of the American buffalo's near-extinction makes clear. The history of settler colonialism in what is presently called the United States is punctuated by violence leading to loss—losses that affect Native peoples acutely, though not exclusively.[3] When the US Army Corps of Engineers and the Bureau of Reclamation united to erect several dams in the Missouri River on and near Oceti Sakowin (Sioux) lands in four present day states—Montana, North Dakota, South Dakota, and Nebraska—the government agencies dispossessed Native inhabitants of more than two hundred thousand acres of reserved lands and inundated "the lush river bottomlands where food and game abounded, subsistence agriculture flourished, and fresh water was plenty."[4] Not only were particular habitats lost in the flood. So, too, were the Oceti Sakowin's ways of dwelling on their homelands. Or, to take another example, when the US government removed them from their homelands in the southern Appalachian mountains, the Cherokee lost access to hundreds of culturally significant plants—plants with which they had learned to coexist and from which they received sustenance and healing.[5] Instances of intentional and unintentional violence against other-than-human beings and ways of being are all too common: damming sits alongside pit mining, stream diversion, overfishing, mountaintop removal, and deforestation, to mention only a handful of the ecologically catastrophic and culturally devastating practices undertaken for the benefit of some humans.

Although some ecological and earthly losses follow from deliberate acts of violence against the land and its more-than-human cohabitants, others are unpredictable—products less of xenophobic malice or ecocidal greed than of ignorance or negligence. The numerous losses that have come, and that will continue to come, as a result of climate change provide a case in point.[6] What climate change makes so painfully clear is that one need not intend to induce a mass extinction event or intentionally devastate habitats or purposefully disrupt earth systems in order for one's actions to generate these effects.[7] Driving to the path where I will hike with my friend, tapping away at my computer's keyboard, or flying to another academic conference, I surely do not intend to add to the losses accumulating as a result of climatic changes—nevertheless, my actions do contribute to the cascade of ecological and earthly losses tied to warming global temperatures, more frequent extreme weather events, catastrophic wildfires, polluted water and air, and reductions in biodiversity.[8] Even well-intentioned efforts to resolve one calamity often open onto other crises.[9] Thus does loss so often beget more loss.

Though loss is an inherent aspect of earthly coexistence which cannot be willed away, there is nothing that guarantees we will notice, much less acknowledge, any particular loss. What accounts for the fact that some of us seem unwilling or unable to sit with ecological losses and earthly transformations? While there are surely many explanations, I want to highlight two potential reasons—one cultural, the other ecological. As will become clear, these two explanations are entangled.

Occasionally, a friend or acquaintance will share a meme I have seen several times before. On one side of the image are basic outlines of six common leaf shapes. Above them, a caption exhorts viewers to "name these plants." Six hand-drawn yet instantly identifiable corporate logos appear on the other side of the image. Viewers are enjoined to "name these brands." Facebook, Nike, McDonald's, Volkswagen, Apple, Lacoste. "This," the message at the bottom of the image announces, "slaps you in the face with reality." Of course, reality according to the meme is partial: for some of us, it is surely easier to name the brands than the plants. For others of us, however, naming the plants will come just as easily as—if not easier than—naming the brands. One's perspective matters, and in the gap between these two experiences, we glimpse the roots of the "cultural" explanation for why some but not all of us notice specific ecological losses and earthly transformations.

To state the obvious, some of us dwell in cultures that alienate us, sensually as well as intellectually, from both the land and the myriad plants and animals with whom we share the land. Obsessed with work, with wealth, with possessions, with mobility, the dominant American culture that I—and, I suspect, many readers of this book—inherited can have this alienating effect. Mainstream American culture is the product of centuries of settler colonialism and, thus, carries on many of its tendencies: the tendency, for example, to see more-than-human beings and ways of being as resources available for extraction, use, or abuse in service of exclusively human (and, worse, only some humans') needs or desires. Or the tendency to treat land as property, something to be bought and sold in the ceaseless flow of capital. Or the tendency to valorize movement and linear progress—the inclination, in other words, to remain enthralled with the novel, with innovation, with a future imagined to be not only different from but also better than the past and present. The more caught up in these tendencies one becomes, the more remote does the land, as well as other animals and plants, come to seem.

By contrast, some of us dwell in cultures which connect us, both intellectually and sensually, to the land and our fellow travelers, cultures which place relationality rather than independence, reciprocity rather than instrumentality, at the forefront of daily life. Despite centuries of settler-colonial violence, for instance, in the present-day United States many Indigenous peoples and worldviews persist alongside and against the dominant American culture I described a moment ago. “The very thing that distinguishes Indigenous peoples from settler societies,” Native American researcher and activist Dina Gilio-Whitaker notes, “is their unbroken connections to ancestral homelands. Their cultures and identities are linked to their original places in ways that define them, as reflected through language, place names, and cosmology or religion.” Indigenous cultures, Gilio-Whitaker adds, tend to view the world as “a world of relationships built on reciprocity, respect, and responsibility, not just between humans but also extending to the entire natural world.”[10] Though my aim here is not to provide a detailed accounting of any worldviews, what ought to be evident from even this brief description is the degree to which Indigenous attitudes toward the land and our fellow travelers differ profoundly from those inculcated in dominant settler cultures: where one worldview stresses separation and utility, the other highlights relationality and responsibility. These differences matter. When relations with particular lands, animals, and plants are central to one’s very being, one cannot help but notice when one or the other is lost or otherwise at stake.

I have no interest, though, in drawing harsh or impermeable lines between cultures or worldviews, as if either could designate something essential, stable, unchanging. It is as true that certain settlers have maintained intimate relations with and cultivated a sense of responsibility to the more-than-human world, for instance, as it is that certain Indigenous people have come to treat the land with disrespect. The myth of the ecological Indian can be just as pernicious as the idea that all settlers and their descendants lack respect for the more-than-human world; we should be wary of taking either narrative as a general truth. Naturalists such as Henry David Thoreau and Aldo Leopold, while clearly supporters and beneficiaries of settler colonialism in certain key senses, nevertheless developed sensuous, intellectual, and moral connections to what exceeds the human, connections which made it possible for them to recognize when more-than-human beings or ways of being were at risk.[11] And, as Gilio-Whitaker points out, a centuries-long history of physical and cultural dispossession in the United States has compromised many Native American practices and principles, leading some Indigenous peoples to engage in extractive activities that threaten to further disrupt their connection to the land and to life- and culture-sustaining traditions.[12] Again, my aim is not to draw hard lines between either cultures or worldviews here; clearly, there are varying experiences that resist reductive explanations.[13] Nonetheless, it behooves us to acknowledge that general differences matter, and that how we understand ourselves in relation to more-than-human others influences our capacity to notice and acknowledge ecological losses and earthly transformations.

While cultural differences partially account for our varying capacities to apprehend such losses and transformations, other forces—ecological forces—are also at play. Here, I take my lead from Kathryn Yusoff, a geographer who has written compellingly about what she calls "the economies of abundance and scarcity." What remains, Yusoff explains, has a way of crowding out what has been lost. What is present has a way of concealing what is absent. "Life as a dynamic and excessive force is often so abundant and present," Yusoff writes, "that it obscures scarcity and literally commands the frame or steps in to the fill the ecological niche."[14] A cultivated field teeming with plants obscures the fact that a verdant forest once stood in its place—a forest and, therefore, an ecosystem that was uprooted. Or the loss of one tree species in a forest prompts the rapid spread of another, thus covering over the prior loss with, literally, new growth. Or the introduction of invasive plants pushes native plants out of their usual habitats, filling the ground with lively matter yet concealing the losses on which their success is premised. There are, of course, other examples.[15] What matters at this point is that the economies of abundance and scarcity, to which Yusoff draws our attention, make it difficult, in many specific cases, to ascertain what has been and what might yet be lost.

Peering out of the south-facing window in my home office, I am greeted by numerous species of birds every day. American robins (*Turdus migratorius*), house sparrows (*Passer domesticus*), blackbirds (*Turdus merula*), downy woodpeckers (*Picoides pubescens*), blue jays (*Cyanocitta cristata*), mourning doves (*Zenaida macroura*), black-capped chickadees (*Poecile atricapillus*), and northern cardinals (*Cardinalis cardinalis*), among others, perch in the eastern hemlocks (*Tsuga canadensis*) and red pines (*Pinus resinosa*) which separate our neighbor's yard from our own. Eastern gray squirrels (*Sciurus carolinensis*) and eastern chipmunks (*Tamias striatus*) scurry up and down these same trees, bounding from branch to branch at dizzying speeds. An occasional Cooper's hawk (*Accipiter cooperii*) swoops in and scatters the smaller birds and mammals. On quiet days, I can hear owls who spend their days in nearby woods, hooting back and forth, back and forth: *hoo hoo-hoo hoo hoo*. Great horned owls (*Bubo virginianus*), I guess. Peering out my window, my attention is drawn to all that is—to all that remains. The sheer abundance of more-than-human beings and ways of being on the other side of the glass has a way of obscuring the transformations to this now-suburban landscape, of concealing the lives and ways of life that have slipped, or been pushed, away in the making of this place and time I now inhabit. The losses easily fade from view.

Fundamentally, this is a book about what it means to become capable of noticing ecological losses; of acknowledging the impressions that these losses leave on ourselves and others, human and more-than-human alike; and of responding to past, present, and potential losses in ways that fortify rather than undermine the conditions of earthly coexistence. Still more precisely, this is a book about some of the ways we are moved—and might move others—not only to notice those more-than-human beings and ways of being that have been or might be lost but also to regard them as worthy of our grief and, thus, of our care.

Animating my work here and elsewhere is the conviction that humans can and should cultivate capacities for attending to, sitting with, and responding to ecological losses and earthly transformations.[16] This conviction emerges out of my encounters with writers, ecologists, artists, and activists who are already quite adept at noticing, grappling with, and seeing the significance of more-than-human loss—and who help others become better able to do the same. I have in mind writers like Linda Hogan (Chickasaw), who teaches readers that, "there are already so many holes in the universe that will never again be filled, and each of them forces us to question why we permitted such loss, such tearing away at the fabric of life, and how we will live with our planet in the future."[17] I have in mind ecologists like Aldo Leopold, who powerfully demonstrates the connections between education and our ability to apprehend losses. "One of the penalties of an ecological education," Leopold once remarked, "is that one lives alone in a world of wounds."[18] I have in mind artists like Olafur Eliasson, who shows us the role that public art plays in heightening our attention to climate change and the losses it entails.[19] And I have in mind activists like Winona LaDuke (Ojibwe), who looks unflinchingly at what has already and might yet be lost, and who then works to build better, less violent worlds.[20]

Common to these and the many other writers, ecologists, artists, and activists with whom I engage in the following pages is a rejection of human exceptionalism, an ideological and philosophical precept that human beings are both inherently different from and better than what is not human. Underwritten by a dualistic and hierarchical image of thought,[21] human exceptionalism defines what it means to be human in opposition to a variety of others, thus drawing powerful, albeit imaginary, dividing lines. Divisions such as human-animal, mind-body, culture-nature, and rational-irrational, each of which implies not just differentiation but subordination, coordinate to sustain what Thom van Dooren calls "the core of a human exceptionalism that holds us apart from the rest of the world and, as such, contributes to our inability to be *affected* by the incredible loss" of what exceeds the human.[22] This is so because human exceptionalism teaches us that we are apart from, rather than a part of, the natural world. It teaches us to treat other living beings and non-living entities as resources there for the taking, as means to anthropocentric ends. Human exceptionalism teaches us, too, that only human lives and deaths matter (and, what is worse, that only some human lives *really* matter), that the loss of plants and animals and habitats and ecosystems is no loss at all so long as some people benefit.

I have written this book *against* human exceptionalism—against the belief that we are separate from and superior to other animals and plants and the lands we cohabit. Human exceptionalism, pervasive though it is within many epistemologies, is not the only show in town. To take just one example, Indigenous ways of knowing demonstrate how else human beings can imagine our embeddedness in and entanglement with the more-than-human. As Potawatomi botanist Robin Wall Kimmerer explains, "One of our responsibilities as human people is to find ways to enter into reciprocity with the more-than-human world."[23] One of our responsibilities, in other words, is to re-weave ourselves into relationships with plants and animals and the land—relationships defined by a mutual give and take, relationships built on a foundation of respect and shared obligations. Aldo Leopold's notion of the "land ethic," centered as it is on a conception of community that includes human and more-than-human alike, also encourages us to enter into respectful and reciprocal relations: "In short," writes Leopold, "a land ethic changes the role of *Homo sapiens* from conqueror of the land-community to plain member and citizen of it." Placing humans within rather than beyond this relational web, the land ethic "implies respect for his fellow-members, and also respect for the community as such."[24] In writing against human exceptionalism, I write *for* modes of relationality, reciprocity, respect, and responsibility that leave us feeling more connected to the earth that bore and sustains us.

Being part of a land community means enduring losses. It means mourning when a member of the community, or the community itself, ceases to exist. Loss, as many writers have pointed out, is the cost of connection. The more connected we feel to others, the more likely we are to feel their loss. And to feel it intensely. Let us call this intensity of feeling bereavement: the sense of deprivation we experience when someone to whom or something to which we are tethered dies, disappears, or otherwise becomes absent. We feel bereaved when a connection has been severed, when a bond has been broken. If we feel disconnected, if we feel there was no bond in the first place, this sense of deprivation is unlikely to set in. But when it does set in, when we do feel that a significant relation has been broken, it can catalyze deeply ethical, political, and practical commitments. As Douglas Burton-Christie explains, "the ability to mourn becomes one of the most important signs of whether, in the face of devastating loss, persons and society as a whole can learn to act on behalf of genuine social, political, and ecological renewal."[25] Put differently, the capacity to mourn beyond the human is linked to the capacity—indeed, the desire and the willingness—to undertake the difficult yet necessary work of maintaining the conditions of earthly coexistence. While we may not be able to avoid ecological and earthly losses, we can certainly respond to them in ways that foster better, less violent relations in the future. Thus cultivating capacities for mourning other animals, plants, and the land itself can lead us out of human exceptionalism and into relations defined, for instance, by "care, consent, and reciprocity."[26]

How, though, do people become capable of mourning ecological losses and earthly transformations? Under what conditions are people able to see the loss of more-than-human beings and ways of being as worthy of our grief? These are some of the questions I take up in this book. In doing so, I ask after the conditions of what Judith Butler calls "grievability," the capacity to see another life or way of life as one that one would grieve and, thus, as one that ought to be safeguarded against loss.[27] Understood as a socially and politically crafted capacity, grievability names not an inborn characteristic but a mode of responsivity that is highly dependent on the spheres within which we find ourselves. In other words, we cannot take for granted this capacity to grieve, but we can hone it individually and collectively, as I hope to demonstrate in the following pages. Beginning in the next chapter, I will argue that our capacity to grieve beyond the human should be understood as an achievement. As a scholar of rhetoric—a concept I describe in the next chapter—I home in on some of the specifically rhetorical conditions of ecological grief, paying special attention to the roles that naming (chapter 2), archiving (chapter 3), and making-visible (chapter 4) might play in preparing us to "mourn beyond the human."[28] To focus on rhetorical conditions and practices in this way is not only to diagnose from whence our capacity to grieve beyond the human originates but also, in some significant sense, to prescribe a course of treatment for the human exceptionalism, and for the fantasy of separation, and for the estrangement that threatens the very conditions of earthly coexistence.

Before considering the relationship between rhetoric and our ability to grieve more-than-human others, though, we would do well to reckon with at least some of the all-too-real biological, ecological, and earthly losses which both animate this work and make grief work so pressing at this juncture.

Day in and day out, plants and animals (including human animals) and fungi come into and pass out of this world. Death is so much a part of life, so constitutive of earthly coexistence, that it tends to recede into the background until it affects us personally—until our beloved family pet dies, until "pests" decimate our favorite trees, until we drive by a deer's bloody body on the highway, until we stumble upon a patch of feathers on the sidewalk, until we kill the spider crawling on our bedsheets. With and without our "help," other living beings are dying, quite literally, all the time. But, of course, humans are responsible for the deaths of a great many more-than-human animals. Most directly, we kill chickens, turkeys, cows, pigs, ducks, sheep, fish, and shellfish for our own consumption. Americans alone slaughter approximately 55 billion animals each year. That amounts to roughly 1,700 animal deaths every second of every hour of every day.[29] Of course, on much smaller scales, we hunt and fish and, therefore, sometimes take the lives of wild animals for food, for fun, or for both. Those animal deaths are intentional, but many animals die for no reason at all. Every day, drivers in the United States run over and kill about one million vertebrates.[30] According to the US Department of Transportation's Federal Highway Administration

2008 report on wildlife–vehicle collisions, "road mortality is among the major threats to the survival" of at least twenty-one animal species that are already considered threatened or endangered.[31] And so, tremendous numbers of individual animals routinely die as a result of both purposeful and accidental human actions.

If we can approximate the number of animal deaths for which humans are directly and indirectly responsible, it is much more difficult to measure the magnitude of plant loss. It is hard to fathom the scale and intensity of vegetal annihilation. Recently I watched one of our neighbors systematically uproot every last dandelion (of the genus *Taraxacum*) from his yard; wielding a small tool that enabled him to dig into the earth and retrieve the plants' deep-diving taproots, which can descend a foot-and-a-half or more into the soil below, he pulled hundreds of the sunny flowers from the ground and piled them high. The dandelion, like so many other constituents of the kingdom Plantae, is often seen as a "weed." Simply put, weeds are out of place. Wild and unwanted, weeds should be destroyed. In fact, "weed" serves as both a noun and a verb. One *weeds* weeds; to weed is to get rid of undesirable plants. Most farmers in the United States rely on industrially manufactured herbicides, including glyphosate (the active ingredient in Roundup), to kill weeds that commingle with crops. Road crews deploy herbicides and fire alike to prevent weeds from growing along the edges of interstates and highways. Weeding is so much the norm in the United States that we scarcely notice its deleterious effects on the individual plants who find themselves in the line of spray or fire. Of course, our impact on vegetal life does not stop at weeds. We also cut trees, for instance, and convert them into lumber, paper, furniture, fuel, and much else. We cut—which is to say, we kill—trees for practical as well as aesthetic reasons; a few trees excised from a hillside can open up a hitherto obscured vista.

My point is not to condemn but rather to indicate the ways in which our own lives are often predicated on or involve us in the destruction of other living beings. Animals and plants and fungi will, of course, die with or without our help. Because we are responsible for so many individual organismic deaths, however, it behooves us to acknowledge the role we play—personally and collectively, consciously and unconsciously—in destroying other living things. The fact that we must eat to live means that we *must* kill. There is no way to be without eating, and no way to eat without killing. So, I stand with Jacques Derrida and others who insist that we must cultivate more ethical ways of killing if we wish to develop more ethical ways of being-on-earth-with others.[32] We can and should look to communities for whom subsistence hunting has been or is a primary means of securing food. Moreover, as we have already seen, many of the deaths humans cause do not support our own survival and flourishing; many serve no "purpose" at all. Part of the task, then, must be to own up to the losses our being necessitates as well as to confront the superfluous losses humans generate. Though we may smash a spider, uproot a weed, or chop down a tree, we often fail to see these actions as acts of destruction—that is, until another living being finds itself on the edge of extinction.

Extinction, like organismic death, is a part of life. Species come and go, evolve and wither away, often over very long periods of time. Of all the species of plants and animals that have ever lived on earth, it is estimated that just one percent remains. For one reason or another, the other 99 percent are now extinct.[33] Most of these extinct species were gone long before humans entered the scene, so what we know of them derives largely from the fossil record. Looking backward, biologists can calculate what is called the "background extinction" rate, an estimate of how many species, under "ordinary" circumstances, can be expected to die out over a particular period of time. The calculations are rough but helpful indicators of how current extinction patterns align with, or differ from, what is expected. For mammals the background extinction rate is perhaps surprisingly low: one mammalian species every seven hundred years is likely to become extinct.[34] As a result, to an ordinary human being, species extinction is practically imperceptible. Individual plants and animals die out in a matter of hours, days, months, years, or decades, but species fade away in slow motion, precariously perched as they are on what ecological philosopher Thom van Dooren describes as the "edge of extinction." Extinction is usually a slow burn, not a spectacle. And so, rather than seeing extinction as a "sharp, singular event," van Dooren suggests we are better served by treating it as "a slow unraveling of intimately entangled ways of life that begins long before the death of the last individual and continues to ripple forward long afterward."[35]

While extinction is as natural as organismic death, there are times when species die out in numbers that far exceed the usual background rate. We now know that earth's history has been punctuated by five prior mass extinction events, the most recent of which occurred some 66 million years ago. In the late 1970s an American geologist named Walter Alvarez discovered a layer of iridium-rich clay in the Gola del Bottaccione, a gorge just outside of Gubbio, Italy. Iridium, a kind of platinum, is rare on earth but is common in meteorites and asteroids. Over the next few years, Alvarez and others developed what came to be known as the "impact hypothesis."[36] As Elizabeth Kolbert explains it,

> On an otherwise ordinary day sixty-five million years ago, an asteroid six miles wide collided with the earth. Exploding on contact, it released energy on the order of a hundred million megatons of TNT, or more than a million of the most powerful H-bombs ever tested. Debris, including iridium from the pulverized asteroid, spread around the globe. Day turned to night, and temperatures plunged. A mass extinction ensued.[37]

Alvarez's insights helped to overthrow the idea that extinction is always a gradual process. Sometimes extinction is rapid—rapid, that is, in a geological sense. The End-Cretaceous extinction event shattered life on earth, disappearing most dinosaurs and sparing only about 20 percent or so of plants and animals. The majority of the flora and fauna with which we are familiar—including *Homo sapiens*—evolved in the wake of the last mass extinction event. Since *Homo sapiens* evolved well after the last mass extinction event, we have yet to encounter the scale of biodiversity loss that the so-called Big Five extinctions wrought on the earth.

This is changing as we move into what some ecologists are describing as the "Sixth Mass Extinction." Unlike the last major extinction event, though, there are no asteroids or other external forces to blame. Rather, for the first time in earth history, *Homo sapiens* bear responsibility for setting a mass extinction event into motion. A group of researchers from the United States and Mexico reported that "our global society has started to destroy species of other organisms at an accelerated rate."[38] They cite population size and growth, and the increase in consumption which follows, as the primary triggers for the acceleration of species-level extinction. In the spring of 2021, as I put the finishing touches on this chapter, roughly 7.6 billion human beings live on earth. Just sixty years ago, the human population was slightly more than 3 billion.[39] In only a half-century, the number of humans has doubled, as has our demand for energy and other finite resources. While only one indicator of our consumption, records of global carbon dioxide emissions yield a telling correlation: in the mid-twentieth century, as the number of humans rapidly grew, so, too, did carbon dioxide emissions begin their upward ascent.[40] Along with rising greenhouse emissions, ever-increasing numbers of people demand so-called "development," which is better understood as a euphemism for habitat destruction and ecosystem disruption: razing forests, draining wetlands, and paving over fields, to name just a few of the ways that human "progress" can make the earth less livable for other species.

Since 1964, the International Union for Conservation of Nature (IUCN)—a group of about 15,000 experts—has assessed more than 116,000 species to determine whether they are at risk of extinction. Of the plants, animals, and fungi the IUCN has evaluated, about 31,000 species find themselves "threatened with extinction, including 41% of amphibians, 34% of conifers, 33% of reef building corals, 25% of mammals, and 14% of birds."[41] And of those more than 6,500 are "critically endangered," meaning that, without significant help, they are very likely to go extinct in the wild. In the southern Unites States, Alabama's plicate rocksnail (*Leptoxis plicata*) has been pushed to the brink of extinction as its native wetland habitat has given way to housing developments and agriculture. The Nassau Grouper (*Epinephelus stratius*), which dwells in parts of the Gulf of Mexico, the Caribbean Sea, and the Atlantic Ocean, is severely threatened by overfishing and loss of habitat (coral reefs) due to climate change. Meanwhile, back on the land, green ash (*Fraxinus pennsylvanica*)—whose range covers much of the central and eastern United States—faces an existential threat in the form of an insect smaller than a penny, the emerald ash borer (*Agrilus planipennis*), which humans have unwittingly and unknowingly dispersed across much of North America. And eastern hemlocks face a similar threat in the form of the hemlock woolly adelgid (*Adelges tsugae*), a tiny aphid-like bug that literally saps nutrients from its host.[42] These are, of course, just four of thousands—and perhaps many more—of species that today find themselves under threat of extinction.

Many species not threatened with extinction face another challenge: the loss of their habitat as a result of local disruptions or shifts caused by climatic changes. When a habitat is destroyed, polluted, or otherwise becomes unlivable, species must either migrate to more suitable conditions or confront population-level loss. Some species are more resilient, and more mobile, than others; those who cannot quickly adjust or relocate struggle to survive under conditions they have not evolved to inhabit. From 2003 to 2013, Richard Primack, a professor of biology at Boston University, carefully studied the flora and fauna in Henry David Thoreau's Concord woods. From 1851 to 1858, Thoreau had himself methodically chronicled more-than-human goings-on in the woods and fields surrounding Walden Pond. "For eight years," Primack explains, "[Thoreau] recorded the flowering times of over three hundred plant species, in addition to many observations of other seasonal events, such as the arrival of migrating birds, the leafing out of shrubs and trees, and ice-out on Walden Pond."[43] More than a century-and-a-half later, Primack and his team of researchers set out to see whether and how climate change and other disruptions have affected biodiversity in Concord. They discovered that a third of the species Thoreau had tracked now exist only in small, relatively isolated populations. What is more, over the ten years they conducted research in Concord, they could not locate a full 25 percent of the species Thoreau had seen in the 1850s. Neither could locals point them to specimens of these species. These species have vanished from areas where they once thrived. They are, in ecological terms, "locally extinct." Extirpated.

Local extinctions are not uncommon. Scientists have documented local extinctions for hundreds of species across a wide range of habitats and ecosystems as a result of climate change and its cascading effects.[44] The reasons for a particular local extinction are complex and difficult to summarize in any general way. However, rising temperatures can set off a chain reaction that renders previously habitable areas unlivable. As temperatures rise, many species migrate to areas of higher elevation and higher latitudes. Chasing conditions under which they have evolved to survive and thrive, these species tend to move up and north to relatively cooler climates. This is called a "cold-edge expansion." Not all species, though, can rapidly adapt to changing conditions or shift their range, and it is these plants, animals, and fungi that are especially at risk of local extinction. If a species cannot adjust or migrate quickly enough, it is vulnerable to a "warm-edge contraction." The ecologist John Wiens explains that warm-edge contractions occur "when populations from one or more localities at the lowest latitudes or elevations of a species' regional distribution disappear."[45] These species, unable to keep up with the pace of climatic changes, wither away under the stress. In Concord, Primack's team of researchers failed to find specimens of many of the plants that Thoreau had seen in the 1850s: the elusive arethusa orchid, six species of bladderwort, and several kinds of mountain mint, to name just a handful, had disappeared.[46] Thoreau himself was awake to the likelihood that at least some of the species he encountered would one day disappear from Concord. "The era of the Wild Apple," he wrote in the *Atlantic Monthly* in 1862, the year of his death, "will soon be past. It is a fruit which will probably become extinct in New England. I fear that he who walks over these fields a century hence will not know the pleasure of knocking off wild apples."[47]

On scales both local and global, biodiversity is waning. And though it is important to acknowledge these losses, we also need to recognize that the diminishment of any given species often leads to the deterioration of others. The first and most important principle of ecology is that all things are interconnected and interdependent. When one thing struggles, we should not be surprised to discover that others do, too. Edward O. Wilson, the Pulitzer Prize–winning biologist, explains: "Biodiversity as a whole forms a shield protecting each of the species that together compose it, ourselves included." We are all better off when we are all better off, in other words. Wilson offers the following example:

> When a century ago the American chestnut, once a dominant tree over much of eastern North America, was reduced to near extinction by an Asian fungal blight, seven moth species whose caterpillars depended on its vegetation vanished, and the last of the passenger pigeons [who depended on chestnuts for food and habitat alike] plunged to extinction. As extinction mounts, biodiversity reaches a tipping point at which the ecosystem collapses.[48]

Wilson's example underscores the importance of safeguarding biodiversity, for the loss of one species may lead to the decline of others. Consider the quintessentially American lawn: a swath of deep-green grass, free of weeds and, especially, of dandelions. Herbicides make the ideal lawn possible, eradicating those species of flora deemed unsightly or impure. Yet, by ridding our yards of dandelions, we also diminish a crucial, early spring food source for honeybees and other pollinator species. And so it is that a seemingly insignificant piece of the ecological puzzle turns out to play an important role in sustaining biodiversity. Without pollinators, countless other plants, including many that we consume, cannot reproduce or bear fruit.

It is not just living organisms and species that we are losing. We are also losing the habitats and ecosystems upon which many species depend. All fungi, plants, and animals require a sustaining set of conditions—peculiar to each species—if they are to survive and thrive. Whenever those conditions are degraded, disrupted, or destroyed, the effects ramify throughout the community. Diminish the conditions themselves and you also diminish the relations that have evolved to flourish thereunder. Again, loss tends to beget loss.

Consider what happens when too many trees are cut down, cut up, and transported away from the earth that bore them. We call this process, rather clinically, "deforestation." When deforestation is wholesale in nature, we call it "clear-cutting." Roughly 17 percent of the Amazon rainforest has been lost over the past half-century, and the World Wildlife Fund estimates that some 18.7 million acres of forests are cut annually.[49] But forests are not just the trees they comprise—or the lumber, fuel, and capital into which they can be converted. Forests are more-than-human communities, teeming with an astonishing variety of living beings and ways of being, many of which can only survive in forest habitats—and, then, only in particular forest habitats. Approximately 80 percent of earth's terrestrial biodiversity dwells in forests,[50] including many species of birds and small mammals that rely on trees for habitat, food, building materials, and protection from storms and predators alike. Wildflowers and shrubby plants grow up in the shade of the tree canopy. Mycelia and fungi make their home on the forest floor, breaking down fallen leaves and rotting trees. And forests are, as we learned when we were young, the earth's "lungs": inhaling carbon dioxide, forests exhale the oxygen upon which we and all other oxygen-dependent beings rely. Without forests, there can be no breathing, no speaking, no thinking. Forests are vital parts of what it means to be human.[51] Among the many "ecosystem services" they provide, forests store tremendous quantities of carbon and, thus, help to mitigate anthropogenic climate change. They are inextricably interwoven with human history and culture.[52] And so, deforestation entails much more than the loss of individual trees. Rather, the losses are simultaneously biological, aesthetic, ecological, and cultural.

Like forests, wetlands provide habitat and food alike to an array of plant and animal species. For instance, coastal wetlands furnish "spawning grounds, nurseries, shelter, and food for finfish, shellfish, birds, and other wildlife."[53] Wetlands filter polluted water, buffer coastlines during storms, and lend stability to shorelines which are shifting due to climate change. But, also like forests, wetlands are exposed to injury and destruction. In their 2013 report, Thomas Dahl and Susan-Marie Stedman describe the net loss of more than 360,000 acres of saltwater and freshwater wetlands in the United States from 2004 to 2009, or approximately one percent of the country's wetlands.[54] Lamentably, these losses are part of a larger trend. In the United States, wetlands have long been viewed as an obstacle to progress and development. At the end of the 1700s, Dahl notes in a report he delivered to the US Congress in 1990, "the area that now constitutes the 50 United States contained an estimated 392 million acres of wetlands." "Over a period of 200 years," Dahl continues, "the lower 48 states lost an estimated 53 percent of their original wetlands," which amounts to losing about "60 acres of wetlands every hour between the 1780's and the 1980's."[55] Such astonishing losses are usually the direct result of human intervention. Wetlands are often drained, backfilled, or otherwise altered to make way for purportedly more "productive" activities: agriculture, grazing, shopping malls, housing developments. And so, the destruction of wetlands often seems more like a gain than a loss: from the soggy, sodden earth, communities quite literally break new ground. But, of course, they lose much more. As the abundance and health of wetlands decline, so too does the profusion and well-being of the myriad plants and animals who dwell within them.

Soil health is also on the decline in many parts of the world. Though soil is usually understood as inert matter, underfoot and out of mind, it makes more sense to see soil as a living thing, susceptible, like all living things, to injury. Where industrial agriculture is or has been most intense, soil health almost inevitably diminishes.[56] This is especially the case when agriculture becomes monocultural. Monocultures sap nutrients from the soil, thus demanding a steady infusion of industrially manufactured fertilizers. Having been depleted from the soil, nitrogen in particular must be added back into the ground to guarantee crop growth. In the United States, just a few crops—corn and soybeans chief among them—dominate our fields, making annual monocultures the rule rather than the exception. And because we tend to cultivate annuals, disturbing the soil is unavoidable, for we must till and sow come springtime. All that ploughing leaves the soil exposed to the elements—to the wind and the rain, both of which send the soil elsewhere.

In America's "breadbasket," the Midwest, much of this displaced soil, along with the millions of tons of fertilizers, herbicides, and pesticides, which farmers increasingly deploy to maximize yields and profits, flows down river and drains into the Gulf of Mexico. As the excess nutrients and chemicals from upstream settle into the Gulf, they catalyze a deadly chain reaction that generates a hypoxic, or "dead," zone where marine life struggles to survive. In June 2019, the US National Oceanic and Atmospheric Administration discovered a "dead zone" more than 7,800 square miles in size in the Gulf of Mexico.[57] It is thus that one problem becomes two: the degradation of the soil gives way to the diminishment of marine life hundreds, even thousands, of miles downstream.

And yet, the accumulation of excess nutrients and industrial chemicals is only one of many threats to the earth's oceans and, thus, those beings and ways of being which dwell within. As the title of one report puts it, oceans are becoming "hot, sour, and breathless."[58] Acidification, warming, and deoxygenation are among the most significant threats that oceans face in an era defined by climate change. In fact, the earth's oceans have helped to mitigate the impacts of global warming by storing some of the excess carbon dioxide we have been emitting. As oceans absorb surplus carbon dioxide, however, they become more acidic. And as pH levels decrease to levels not seen in more than 60 million years, many species are suffering. In particular, species with calcium carbonate skeletons and shells—corals, mussels, oysters, and the like—are experiencing the brunt of these changes to their habitat. Like acidification, warming waters directly impact ocean-dwelling species. "There has already been a mean sea surface warming of about 0.7°C over the last 100 years, likely to increase by over 3°C in some ocean regions by the end of this century."[59] As the waters warm, some processes slow down and become less frequent. Nutrient mixing, which brings nutrients stored in deeper waters up into the relatively nutrient-poor surface waters, tends to occur less often in warmer waters. When the nutrient content of surface water declines, it becomes less livable for marine life. Warming is also linked to deoxygenation. As a rule, warmer water contains less oxygen than colder water. Oxygen levels can also be lowered, as I noted, by chemical and nutrient run-off from upstream, as is happening in the Gulf of Mexico and elsewhere. It perhaps goes without saying that lower oxygen levels directly impact marine life. Organisms that require more oxygen than their environment can supply die or migrate. At the same time, though, those organisms that can tolerate a less oxygen-dense habitat are likely to thrive in the warmer, deoxygenated water, further disrupting the marine community and displacing endemic species.

Forests, wetlands, soils, and oceans are experiencing tremendous diminishments, but the cryosphere is playing host to perhaps the most iconic ecological losses of our time—rapid glacial melting, sea ice loss, and permafrost thawing. As global temperatures rise, the ice melts and releases liquid water and any chemical compounds, including carbon dioxide and methane, that have been stored in the ice. The second largest body of ice on earth, for instance, the Greenland ice sheet, has been losing mass every year for more than 20 years. As the water stored in Greenland's glaciers melts, sea levels rise. According to a 2019 report in the *Proceedings of the National Academy of Sciences*, melting in Greenland "has raised sea level by 13.7 [millimeters] since 1972, half during the last 8 years."[60] Like glaciers, sea ice—frozen seawater that floats on the surface of the ocean—is also withering away. Sea ice forms and retreats each year; as it does so, it regulates water temperature, encourages nutrient mixing, and provides habitat and hunting grounds for a number of species, including seals and polar bears. The Arctic, where sea ice coverage has been declining "in every geographic area, in every month, and every season" for decades, "may be ice-free in summer before the end of this century."[61] The Arctic's permafrost is also diminishing. Less discussed than glacial melting and sea ice loss, thawing permafrost—ground that remains frozen—also threatens to unravel ecosystems and release potentially massive quantities of methane, nitrous oxide, and carbon dioxide into the atmosphere. As permafrost thaws out, the landscape shifts. Ground that once was frozen and hard becomes mobile, "triggering landslides and slumping at alarming rates, resulting in stream flows changing, lakes suddenly draining, seashores collapsing, and water chemistry being altered in ways that could be deleterious to both humans and wildlife."[62]

On and on this list of horrors could go.

While only a partial accounting, this litany of ecological losses and earthly transformations suggests some of what is at stake in negotiating coexistence. It also suggests the magnitude of the task we face as we shore up existing relationships with more-than-human others and as we establish new connections with what exceeds the human. One consequence of engaging in this work, as I have said before, is becoming more aware of all that we have lost, all that we are losing, and all that we might yet lose. And so, sadness is inevitable. We grieve when we feel that a connection has been broken, a relationship dissolved. We need to grieve, for in doing so we acknowledge the other beings and ways of being to whom and to which we are linked. This grief—this acknowledgment of loss—can open onto healing.

To heal ourselves and our relations with other plants and animals, with the land and the earth, we need more than knowledge of environmental crises. We need, too, to feel the weight of losses past, present, and to come. This challenge is not new. Indigenous peoples, dispossessed of their ancestral and traditional lands and severed from sustaining relations with animals and plants, have for centuries known and felt the immensity of such losses.[63] This convergence of knowledge and feeling, of awareness and emotion, is necessary for an ethical orientation to the more-than-human to be generated and sustained. As Aldo Leopold writes in "The Land Ethic," "No important change in ethics was ever accomplished without an internal change in our intellectual emphasis, loyalties, affections, and convictions."[64] It is in this spirit that I have written this book: in the spirit of lighting up other emphases and loyalties and affections and convictions. In the spirit, that is, of healing.

Grief can be a path to healing, a "path to understanding entangled shared living and dying."[65] Painful though it may be, there is hope in becoming mournful. "Emotions matter because they orient us toward the world," Jeremy Engels writes. In orienting us, feelings affect how we see ourselves and others, such that we "see the world differently depending upon our emotional state."[66] How do we see the world when we are feeling mournful? How might cultivating capacities to grieve beyond the human alter our relationships with more-than-human others?

Mournful, we come to see the earth as a scene of relations. Mournful, we come to value attachment and connection across difference. Mournful, we come to see these attachments, these relationships, as worthy of protection and care. Let us turn, then, to a consideration of how we become capable of grieving beyond the human.

CHAPTER 1

Ecological Grief

A Rhetorical Achievement

Feeling Ecological

Although this is a book primarily about ecological grief, I presuppose that ecological grief is just one among several feelings that lend shape and texture to our relationships with what exceeds the human. Awe, fear, disgust, enchantment, wonder, sympathy, empathy, anxiety, despair, worry, depression, denial, carelessness, guilt, shame, indifference, apathy, delight, rage, sorrow, desperation, gratitude, curiosity . . . the list goes on and on. Our feelings for, about, and in response to the more-than-human world are as complicated—as beautiful and as heartbreaking—as the vexed and vexing period of earth history we now inhabit.

I also presuppose that some feelings are more conducive to earthly coexistence than others, that some feelings motivate us to care for other animals and plants, species, places, land communities, and earth systems in ways that other feelings do not. In *Earth Emotions*, environmental philosopher Glenn Albrecht describes two emotional forces that influence humans' relations with the earth: what he calls terraphthoric and terranascient feelings. On the one hand, terraphthoric feelings lead to destruction. From the Latin noun *terra*, meaning "land" or "earth," and the Greek *phthorá*, meaning "destruction," terraphthoric emotions are, literally, earth-destroying. Avarice, for instance, may move us to degrade ecosystems as we pursue our own material comfort, wealth, or power. On the other hand, terranascient feelings nurture earthly life. Replacing the Greek *phthorá* with the Latin *nātūra*, meaning "birth," terranascient emotions drive us to safeguard the earth's ability to give birth, again and again, to the conditions that every living being needs to survive and flourish. Love, for example, might move us to enact some restraint—to prohibit ourselves from engaging in some activities so that others might dwell well.[1]

If some emotions encourage us to participate in destructive deeds and others compel us to care, then the prospects of earthly coexistence hinge to a certain extent on our feelings. Our feelings matter because they affect the way we encounter the world. When we feel fearful, the world seems frightful. When we feel bored, the world around us appears boring. When we feel enchanted, the world strikes us as mysterious and intriguing. Our feelings matter, in other words, precisely because they make the world itself matter to us in some ways rather than others.[2]

Given the substantial role that feelings play in shaping earthly coexistence, it should come as no surprise that scholars in the interdisciplinary environmental humanities devote considerable attention to the formation, circulation, and ramifications of particular feelings, from anxiety to apathy, care to carelessness, denial to despair, guilt to gratitude, longing to love, and many others besides.[3] Detailed studies of ecological feelings tend to underscore two key points. First, awareness of or knowledge about environmental and planetary crises alone does not guarantee that one will work to mitigate the disasters about which one has learned. Knowledge is an important yet insufficient condition for earthly care, as numerous scholars have demonstrated.[4] Second, different feelings move us in different directions—toward distinctive values, thoughts, and inclinations. For instance, when we feel personally guilty for triggering or exacerbating ecological calamities, we may come to see ourselves, rather than the economic, political, social, and cultural behemoths into which we were born, as responsible for solving earth's many problems. This is a recipe for disaster, resignation, and self-loathing since the earth's enormous challenges can only truly be solved through collective action.[5] Setting this specific example aside, recent considerations of ecological feelings across the environmental humanities underscore the value of understanding where our earth-destroying and earth-sustaining emotions come from, as well as how they orient us to the more-than-human in ways that make earthly coexistence more or less possible. In this book, I carry that project forward.

Lest readers detect a too simple determinism in the picture of ecological feelings I have been painting thus far, I hasten to add that even terranascient feelings such as love are complicated and open up an array of ethical and practical quandaries. If we are inclined to protect what we love, it matters not just that we love but who and what, in particular, we love.[6] Do we, for instance, love the whole or only specific parts of the whole? This question assumes special significance in connection to matters of earthly coexistence, for the myopic defense of any given part—say, a particular species—may well threaten the integrity of the whole.[7] Love of a native species has led many to wantonly destroy others deemed "exotic" or "invasive."[8] It is also true that our affection for endangered species often commits us to enacting certain kinds of violence in the name of conservation: containment, for instance, and forced impregnation.[9] My point here is not to stimulate suspicion about terranascient feelings so much as it is to note that even feelings like love deliver us over to a wide range of beliefs, values, convictions, and behaviors. Thus, while it is possible to demonstrate how particular feelings orient us toward others, it is important to avoid deterministic claims that posit a simplistic cause–effect relation between feelings and deeds. Our ecological feelings surely influence how we perceive, think about, and respond to the more-than-human world, but they do not therefore lead straightaway to easily predictable outcomes.[10]

Despite the ambiguities and complexities of our ecological feelings, and even though we can never say with certainty that love—or hope or awe or empathy—will lead us to make the "right" decisions or that guilt will suppress our sense of efficacy, we should still investigate, describe, and analyze such feelings for at least three reasons. First, by naming our ecological feelings we make them more accessible. It is often said that until we name emotions it can be difficult to "see" them operating in our lives. Furthermore, terms like "eco-anxiety" and "climate grief" validate marginalized feelings. Second, by describing our ecological feelings we can better understand how they variously disclose aspects of the world to us, as well as how they invite us to imagine and enact our own agency. Only when we have a sense of how ecological feelings orient us in the world and toward others can we determine which to cultivate and which to discourage. Our goal, I hope it is clear, should not be to subject our feelings to the domesticating tyranny of a *logos* hellbent on mastering and subduing *pathos*. The aim is not to master our feelings by getting rid of them. Instead, describing ecological feelings renders us more emotionally intelligent and better equipped to put our feelings to work.[11] Third, and relatedly, by tracing the conditions that make specific ecological feelings more likely to emerge we shine a light on the tactics by which we might individually and collectively cultivate those feelings that incline us to care about and care for the earth and our collaborators in coexistence. To that end, understanding what sets our ecological grief into motion further enables us to engage in those activities which catalyze and sustain our terranascient feelings.

Ecological Grief

Grief and mourning are often distinguished. Whereas grief is situated as an inward, private, emotional experience, mourning is framed as the outward, public expression of one's inner feelings.[12] According to this distinction, grief is the natural emotional response registered within an individual's subjective experience of loss. In grief we recognize, both cognitively and emotionally, that a loss occurred. By contrast, mourning consists of various culturally specific practices, rituals, ceremonies, and other activities through which we adjust to the new conditions created by the loss. Not only are grief and mourning distinguished by their internal and external natures. According to this distinction, grief precedes mourning: before we mourn, we must experience grief. Indeed, mourning is often framed as the way in which we resolve or work through our grief and, thus, get over or move past the loss.

While this distinction between grief and mourning can be useful, the line between grief and mourning often blurs; in everyday speech, for instance, we regularly use "grief" and "mourning" as synonyms. There is some wisdom, I believe, in this terministic slippage, which troubles both the internal/external distinction described above and the assumption that grief is prior to mourning. I regard grief and mourning as emotional, ethical, and political capacities that must be cultivated. In this sense, I broadly take my lead from Ann Cvetkovich's notion of "public feelings," feelings which emerge out of and are sustained by certain forms of social, political, economic, and rhetorical life.[13] Emphasizing the conditioned quality of particular feelings—emphasizing, that is, the fact that certain feelings become more possible under some conditions than others—this concept of "public feelings" invites us to consider the forces that conspire to make particular feelings (less) possible or (less) likely. In doing so, we come to see specific feelings as achievements rather than inevitabilities.[14] Seeing the capacity to grieve as an achievement troubles the presumed temporal priority of grief: for instance, we may become more capable of grieving beyond the human through encounters with mourning practices and mournful rhetorics. But I have already gotten ahead of myself. Before considering the relation between rhetoric and our capacity to grieve beyond the human, we should contemplate how others have defined and explored ecological grief and mourning.

Throughout the past two decades, the terms "environmental grief," "climate grief," and "ecological grief" have entered activist, journalistic, scholarly, Indigenous, and clinical discourses related to ecological loss. In 2004, Kriss Kevorkian submitted a dissertation on "environmental grief," which she views as the "reaction stemming from the environmental loss of ecosystems caused by natural or man-made events."[15] Writing more than a decade later in the scientific journal *Nature Climate Change*, researchers Ashlee Cunsolo and Neville Ellis offered what has become the standard definition of ecological grief, which, they explain, is "the grief felt in relation to experienced or anticipated ecological losses, including the loss of species, ecosystems, and meaningful landscapes due to acute or chronic environmental change." Ecological grief is, they hasten to add, a "natural response to ecological losses."[16] These researchers did not invent the feeling of ecological grief so much as they gave it a name and described some of its dimensions.

That said, academic considerations of ecological grief have catalyzed conversations in public forums as well. Today, coverage of ecological and climate grief appears regularly in both environmental and mainstream news media. "Addressing Climate Grief Makes You a Badass," one *High Country News* headline exclaims, "Not a Snowflake."[17] A *CBC News* story describes the toll ecological grief is taking on some Canadian citizens—for one "the grief was [reportedly] so extreme that it became debilitating."[18] Meanwhile, the *Guardian* declares: "Greenland Residents [Have Been] Traumatised by [the] Climate Emergency."[19] The magazine *Scientific American* extolls the importance of "Facing Down 'Environmental Grief'" in one story and promises its readers that "Taking Action Can Cure Your Climate Grief" in another.[20] *Grist*, an environmental news magazine, assures its readers that, "If You're Suffering From Climate Grief, You're Not Alone."[21] The more mainstream *ABC News* soberly intones: "Climate Grief Expected to be Widespread Soon but It's Still Not Openly Acknowledged."[22] Thus has *Undark*, a digital magazine dealing with the relationship between science and social issues, exhorted: "It's Time to Talk about Ecological Grief."[23] Indeed.

We cannot understand ecological grief without understanding loss. Under particular conditions, a loss may elicit grief because it leaves us feeling bereaved. Literally, to be bereaved means to be deprived of something significant—to find oneself lacking what one once took for granted. The death of a parent or grandparent, sibling or child, partner or friend, companion or colleague: such events deprive us of the loving connections and nurturing relationships that these individuals' presence made possible. Their absence diminishes our lives, leaving us, as the applied philosopher Thomas Attig explains, "to relearn virtually all components of our worlds as we experience them."[24] Lacking the support, intimacy, love, responsibility, and routine such others brought to our lives, we are forced to come to terms with—indeed, to cope with—the fact that the absence of a significant other fundamentally transforms our way of being in the world with others.[25] Of course, we lose more than other people, and death is not the only event that deprives us of meaningful attachments. We lose jobs, homes, and money; we lose objects and ideals; we lose friendships and futures. Thus, while death may be the sort of loss with which we are most familiar, losses of many kinds other than death profoundly affect us and alter our lives.[26]

Theorists of ecological grief, including myself, presume that the diminishment and disappearance of other living beings, places, land communities, and earth systems can also leave us feeling bereft. It is not difficult to understand why such losses might deeply affect us, for we are all—regardless of whether we realize or appreciate it—intricately bound up with what exceeds the human, inextricably enmeshed in land communities and ecosystems, reliant upon more-than-human others. Recognized or not, valued or not, these connections, which literally make our lives possible, can be abandoned, severed, destroyed. The air that we breathe can be fouled; the water that we drink can be polluted or poisoned; the land on which we grow our food and build our homes can be contaminated or taken away from us. And as the conditions upon which our health, well-being, and happiness depend deteriorate, bereavement surely seems an appropriate response. Many other kinds of ecological losses leave us bereft and grieving, including the death of a companion animal,[27] the extinction of other species,[28] the extirpation of a species,[29] the pollution of a watershed,[30] the gradual diminishment of an ecosystem,[31] or the loss of land and the cultural traditions and ways of life that tether us to particular landscapes.[32] Precisely because we become human "only in contact, and conviviality, with what is not human," any ecological loss is—theoretically, at least—capable of leaving us feeling deprived of a significant connection.[33]

Of course, not every ecological loss solicits a sense of bereavement. Nor does every ecological loss elicit grief. Whether we grieve the deaths of other animals and plants, the extinction of other species, the transformation of landscapes, the destruction of ecosystems, or the disruption of earth systems depends on whether we feel connected to these more-than-human others in the first place. Just because we *are* related to such others does not mean that we *feel* connected to them. An ontological attachment does not always—or even usually—entail an emotional one. Absent the latter, we struggle to recognize a loss as a loss.[34]

Since some groups of people seem to be especially susceptible to ecological grief, we can learn a lot about what sets ecological grief into motion by pondering why particular people might be more likely to experience these feelings than others. In their pathbreaking empirical work, Cunsolo and Ellis explain that people whose personal, work, and cultural lives continue to bring them into intimate contact with the natural world tend to be strongly affected by ecological losses.[35] This makes sense. After all, the more involved we are with others, the more likely we will be to notice their loss and to feel deprived by their absence. By comparison, the deaths of total strangers rarely affect us. When we feel close to another person, however, their absence often disturbs us profoundly. Intimacy begets bereavement. Indigenous peoples, farmers, and environmental scientists are among those groups whose day-to-day practices bring them into intimate contact with the more-than-human world and who have historically found—and who increasingly find—themselves grappling with the emotional consequences of ongoing ecological and earthly crises.

Having survived centuries of dispossession and displacement at the hands of settler colonists, Indigenous peoples in the present-day United States and elsewhere are familiar with grief. As Winona LaDuke (Ojibwe) notes, "Most Indigenous cultures of the western hemisphere suffer from a historical unresolved grief . . . accumulated over generations of trauma."[36] Characterized as "savage" or "wild" by settlers, as too close the land, Indigenous people in the United States have been forced off their ancestral and traditional homelands; purposely sickened and starved; relocated onto reservations, sent away to boarding schools and unfamiliar urban areas; deprived of life-sustaining connections with animals and plants (understood as medicine), with landscapes, with their human ancestors' bones and spirits; and severed from distinctive ways of knowing and being-with-others.[37] The weight of these traumas surely adds up, leading some to suggest that Indigenous people may experience a distinctive type of ecological grief, one that is entrenched in longer, more complex histories of environmental, cultural, and political loss.[38] For Indigenous peoples, culture and ecology are inseparable: a loss in one area entails the diminishment of the other. Hence, Indigenous ecological grief may be bound up with a coexistential sense that one has lost, is losing, or might lose not just something significant but something *vital* to one's way of being-in-the-world-with-others.

Consider, for instance, the loss of the buffalo (*Bison bison*) on the Great Plains. At the beginning of the nineteenth century, an estimated 40 million buffalo roamed the plains, making bison "the most numerous grazing animals on earth" at the time.[39] Throughout the same century, however, settlers nearly pushed the buffalo over the edge of extinction in an attempt, first, to dominate the buffalo market and, eventually, to damage Native Americans who depended on the buffalo.[40] "An attack on the land and the buffalo," writes Nick Estes (Kul Wicasa), "was an attack on Indigenous subsistence practices" and, thus, on an entire way of being.[41] The loss of the buffalo profoundly imperiled—physically and spiritually—Indigenous people across the Great Plains: not only their food source but their identity was threatened.[42] Of course, the buffalo is but one of many examples.[43] And lest readers think that such examples are consigned to the past, consider the Inuit people of northern Canada, for whom climate change is not abstract but all too real. For the Inuit people, climate change is generating a radically altered landscape in which sea ice, which has historically enabled travel, hunting, food storage, and other culturally specific practices, is disappearing before their eyes; as the ice goes, so too does the Inuit's ability to practice traditional ways of life and identities dissolve. Indeed, climate change threatens not only the earth's cryosphere; it threatens the cultures that have developed among the ice.[44] As one Inuit poignantly put it, "Inuit are people of the sea ice. If there is no more sea ice, how can we be people of the sea ice?"[45]

In places where farmlands have been or are being altered through industrialization, urbanization, climate change, or other forces, farmers may experience ecological grief as a result of their links to and labor with the land. In rural Australia's wheat belt, for example, chronic drought has led to crop failures and severe erosion, which has in turn forced many farmers to abandon their fields and, with them, the ways of being they once nourished. Like Indigenous communities, small farmers are often intimately attached to specific places; the farmland becomes a scene of relationships among human and more-than-human actors. To lose one's land, or to witness the deterioration of one's fields, or to be compelled to pursue nonagricultural vocations—such experiences can have profound implications not only on one's economic well-being but also on one's sense of identity. One farmer explained to Ashlee Cunsolo and Neville Ellis that losing their family's farm "would be like a death"; they also affirmed that "there would be a grieving process because the farm embodies everything the family farm is. I think if we were to lose it," the farmer continued, "it would be like losing a person . . . but it would be sadder than losing a person."[46] It would be sadder than losing a person, we can surmise, because losing the farm would mean losing not just "property" but a locus of relations, a place in which many beings and ways of being coexist. For this farmer to lose their farm, in other words, would be for them to lose their place in the world among others—and, thus, to lose a manner of dwelling.

Like Indigenous communities and farmers who retain close ties to particular places, environmental scientists often find themselves on the frontlines of ecological and earthly devastation and, thus, vulnerable to the experience of ecological grief. Numerous scientists have documented their grief as they witnessed species become extinct, biodiversity decline, and ecosystems unravel. In *A Sand County Almanac*, for instance, ecologist Aldo Leopold laments the destruction of a cutleaf Silphium plant, the disturbance of marshland ecologies, the extinction of the passenger pigeon, the death of a single wolf, and much else besides.[47] Leopold understood his grief as something of an occupational hazard: "One of the penalties of an ecological education," he once wrote, "is that one lives alone in a world of wounds. Much of the damage inflicted on land is quite invisible to laymen."[48] When we understand ecology, Leopold reasoned, we are better equipped to perceive when things go awry, when species go missing, or when land communities become unhealthy. Decades later, another ecologist echoed Leopold's sentiment in the pages of *BioScience*. Recounting a moment early in her career when she was researching dogwood trees threatened by the aptly named fungus *Discula destructiva*, Phyllis Windle remembers being overcome by sadness by the news and firsthand encounters with the dogwood anthracnose. "Ecologists," Windle writes, "are both blessed and cursed with seeing natural systems clearly. Thus, we see what is there and also know what is gone."[49] Still more recently, ecologists at Harvard Forest published a deeply mournful collection of essays devoted to the eastern hemlock (*Tsuga canadensis*), which is currently succumbing to the hemlock woolly adelgid (*Tsugae adelges*), an aphid-like insect native to Asia but invasive and destructive in the eastern United States and Canada. Their research, they explain, brings them into regular contact with the "heart-wrenching" demise of a "forest giant."[50]

That many environmental scientists endure ecological grief comes as little surprise. Their work in the field and in the lab can bring them into close physical and epistemological contact with living beings and ways of being, many of whom—and which—are themselves under threat of erasure. That scientists express their sorrow candidly and publicly perhaps does come as something of a surprise, though. In the Western scientific tradition, scientists are not supposed to convey their emotions. They are, as Lesley Head and Theresa Harada explain, expected to keep "the heart a long way from the brain."[51] Despite the occupational expectation to be dispassionate, ecologists like Leopold and Windle demonstrate the value of keeping "the heart" and "the brain" tethered. In doing so, they approach something like Potawatomi botanist Robin Wall Kimmerer's intertwining of reason and emotion, analysis and affection, in *Braiding Sweetgrass*—a work in which Kimmerer braids traditional ways of knowing with Western science, and in which she confirms that sorrow is an appropriate response as one investigates imperiled more-than-human relationships.[52] Three biologists recently reaffirmed not only the value but the necessity of allowing scientists to feel and, crucially, to express their ecological grief. Writing in *Science*, the biologists contend that, "grief and post-traumatic recovery can [actually] strengthen resolve and inspire scientific creativity."[53] Far from weakening it, grappling with grief makes science more honest and leaves scientists better able to deal with pressing ecological and earthly crises.

Still, the fact that environmental scientists apparently feel the need to advocate for their ability to acknowledge and deal with their feelings of bereavement and grief gestures toward a more general repression. Within the dominant culture of the United States, ecological grief remains "disenfranchised" and "socially marginalized," out of sight and out of mind.[54] This is not the case when we lose a beloved person. When another person dies, we usually feel free to express our grief. We sit shiva and attend funerals; we give eulogies and write obituaries; we dress in ways that convey our sorrow and share meals in commiseration and celebration. When animals or plants perish, when species become extinct, or when a land community is fractured, however, we have few traditions or rituals to guide us, few outlets for joining with others in our shared suffering.[55] As a result, even when we are assailed by an intimation of ecological grief, we may well struggle to identify and grapple with these feelings. As Stef Craps notes in his introduction to a special issue of the journal *American Imago* devoted to ecological grief, a great many of us are "somewhat at a loss as to how to adequately navigate the emotional terrain of environmental breakdown."[56]

Mental health researchers and practitioners are interested in precisely this question of how we can "adequately navigate" experiences of ecological loss and grief. Increasingly, mental health experts acknowledge that humans' mental—not to mention physical—health is linked in significant ways to ecological health. The American Psychological Association (APA), to take just one example, now recognizes the influence of climate change on human mental health. In 2014, the APA and ecoAmerica released a report entitled "Beyond Storms and Droughts: The Psychological Impacts of Climate Change." Alongside trauma, anxiety, stress, depression, and hopelessness, the authors highlight grief and a sense of loss as some of the "key impacts of climate change on mental health."[57] One of their tips for helping individuals and communities to grapple with environmental losses is to publicly recognize the role that emotions play in our responses. Doing so, they contend, "can help dismantle paralyzing defense mechanisms" and help people grasp that "being anxious, depressed, or otherwise emotional is a normal response" rather than "something to fight against, ignore, downplay, or suppress."[58] In 2017, the APA and ecoAmerica reissued the report under the title "Mental Health and Our Changing Climate." Substantially expanded, the latest report emphasizes the importance of, among other things, building and sustaining connections to places and cultural traditions.[59] Less formally, community members are finding ways to acknowledge and work through ecological grief together: some college professors now offer courses devoted to understanding experiences of ecological grief,[60] for instance, while others have formed "good grief" networks to support one another through ecological losses.[61]

To summarize: While the terms "environmental grief" and "ecological grief" arose only in the past two decades, the individual and collective experiences they name have long been a part of the human condition. Precisely because we are bound up and entangled with the more-than-human, we are vulnerable to feelings of grief in the wake—and in the face—of ecological transformations and losses. The sheer fact that we are constituted in and by our relations with other beings, species, landscapes, habitats, and earth systems, however, does not ensure that we will notice or grieve their loss, just as our shared humanity cannot guarantee that we will recognize, much less grieve, the loss of another human being. And so, while grief is surely an *appropriate* response to ecological loss, as Lisa Kretz contends, it is not an *inevitable* response.[62] Although Indigenous peoples, farmers, and environmental scientists may be especially susceptible to ecological grief due to historical and/or ongoing connections to more-than-human beings and ways of being, as well as landscapes and places, people who lack similar relationships may be less prone to grieve beyond the human. Certain lived experiences may dull one's sensitivity to ecological losses and earthly transformations, making it difficult to recognize more-than-human losses as losses. And if one does experience a pang of ecological grief, one may discover that the culture they have inherited provides them with few resources or opportunities to genuinely recognize or tarry with their sorrow. Nevertheless, if mental health professionals, academics, journalists, and environmental activists are to be believed, and I believe they are, then grieving beyond the human appears to be on the rise.[63] Paradoxical though it may at first sound, and for reasons I shall describe in the following section, we should consider this a very good thing.

A Heartsickness That Heals

The English verb *to grieve* derives from the classical Latin *gravāre*, which has its roots in the adjective *gravis*, meaning "heavy." The loss of a significant other literally weighs upon us. In their absence, we experience a heaviness of heart that often translates into a physical sluggishness and incapacity. We may slouch into the couch or spread out on the bed, curl into the fetal position or stare into the television. Psychologically, we may struggle to focus or to feign normalcy as the heft of the loss bears down upon our psychic economy. Somewhat ironically, losing someone or something we care about encumbers us with feelings, thoughts, memories, and desires which, while latent all along, suddenly press upon us in new and often unexpected ways. Grieving requires some heavy lifting.

Whether we are grieving the death of another human or the extinction of another species, grief is often intense and painful. In perhaps the best-known popular treatment of grief, Elizabeth Kübler-Ross—who described the "Five Stages of Grief" idea, which moves from denial to acceptance—and

David Kessler characterize grief as "the intense emotional response to the pain of a loss."[64] This pain derives, as I noted earlier, from the breaking of an attachment, the severing of a relational bond. Sometimes, these wounds fester; at other times, they heal. Either way, a tender scar remains to remind us again and again, usually when we least expect it, both that and what we have lost. Tellingly, in the diary he penned following his beloved *maman*'s death, the French thinker and author Roland Barthes shifts effortlessly between "mourning" and "suffering," two ways of naming one experience.[65] Undone by the other's undoing, we sorrow and struggle in their absence.[66] Loss hurts. Grief hurts.

Precisely because it is painful and unpredictable, the dominant culture often greets grief with suspicion, treating it as an impediment to—rather than as a precondition for—ethical and political action. Owing to its agonizing intensity, we are told, grief gets in the way of our practical efforts, impairs our reasoning, and distracts us from the most important work. The labor organizer Joe Hill captured this sentiment well when he exhorted, "Don't mourn, organize!"[67] So did activist and critic Douglas Crimp who, in an essay on the early days of the AIDS epidemic, recalled the feeling among many activists that mourning seemed "indulgent, sentimental, defeatist."[68] Grieving in the midst of an ongoing crisis can feel to some like capitulation or like giving in to loss. When we begin grieving, some worry, we may be tempted to give up the fight. Even scholars who otherwise appreciate the role that feelings play in catalyzing and sustaining the work of earthly coexistence sometimes inadvertently reify this idea. For example, early in *Field Guide to Climate Anxiety*, Sarah Jaquette Ray proclaims that "eco-grief" is "not good for us or for the planet."[69] Fortunately, Ray later fleshes out her position, acknowledging that "facing death" and "embracing grief" can actually lead us to "the love that is necessary in order to sustain the work of climate action,"[70] but her oscillation among these two perspectives nevertheless exposes a presumed tension between grief and action, between mourning and mobilizing for earthly coexistence. Even if unintentionally, pitting grief against action reinforces a notion of grief as individualizing, passive, and apolitical. This is neither the only nor the best way to conceptualize grief.

Rather than assuming that grief turns us back upon ourselves, makes us docile, and summons us away from the political sphere of action, we are better served by recognizing how ecological grief gestures toward a sense of connection, intimacy, and even love across often profound differences. Instead of treating grief as an illness to be quickly cured, we ought to see it as a symptom of *philia*—a painful symptom, to be sure, but one that registers our appreciation of, as Thom van Dooren puts it, "all the ways we are at stake in one another, all the ways that we share a world."[71] When we grieve the loss of another animal, we affirm that our lives are intertwined. When we grieve the extinction of another species, we accept that our own capacities for becoming are diminished in a world populated by fewer ways of being. When we grieve for fractured land communities and disrupted earth systems, we avow the simple yet profound fact that our own lives are dependent on these conditions—that we are not now, nor have we ever been, independent beings. Such affirmations do not spell the end of ecological action, as some worry, but, rather, its beginning.[72]

Regrettably, some of the most enduring ideas about the proper course of grief tend to obscure the ethical and political potential latent within grief. One source for these ideas is psychoanalyst Sigmund Freud's article on "Mourning and Melancholia," which was first published in 1917. Mourning and melancholia are related, Freud explained, yet distinct reactions to the loss of another person, object, or ideal. Although both responses entail "a profoundly painful dejection, abrogation of interest in the outside world, loss of the capacity to love, [and the] inhibition of all activity," melancholia also involves "a lowering of the self-regarding feelings to a degree that finds utterance in self-reproaches and self-revilings, and [ultimately] culminates in a delusional expectation of punishment."[73] This final symptom of melancholia makes it especially dangerous, for in the most severe cases, the melancholic's "expectation of punishment" manifests in acts of self-harm. Thus, what distinguished mourning from melancholia for Freud—at least in the 1917 article—was the former's inclination toward healing and the latter's destructive impulse.

I, too, think that grief can be restorative, but my reasons differ from Freud's reasons in "Mourning and Melancholia." Freud characterizes "the work which mourning performs" in this way: "The testing of reality, having shown that the loved object no longer exists, requires forthwith that all the libido shall be withdrawn from its attachments to this object." This process plays itself out in the mourner's psyche, where memories of the lost other are "brought up and hyper-cathected." By focusing intensely upon such memories, the mourner comes to accept that who or what has been lost no longer exists. Upon realizing this, Freud explained, the mourner relinquishes their affection and attention for the lost other. Herein lies mourning's healing potential, Freud believed: by withdrawing from our attachments to the dead and disappeared, we become free to establish new bonds with the living and with what remains. As Freud famously put it, "when the work of mourning is completed the ego becomes free and uninhibited again."[74] Mourning heals, in other words, because it liberates us, lightens our psychic load, and frees us up to move on in the wake of a loss.[75]

By contrast, Freud initially described melancholia as an impediment to healing after a loss. Whereas his mourner moves on, Freud's melancholic gets stuck. Mired in relations with the dead and disappeared, unable to sever the ties that bind, Freud's melancholic fails to get over the loss. They cannot move on precisely because, having established what Freud called an "identification of the ego with the abandoned object," they refuse to accept reality and concede that anyone or anything was lost.[76] One consequence is that "the love-relation need not be given up."[77] This refusal enables the melancholic to maintain their relationship with the lost other, yet it simultaneously prevents them from forming healthy attachments to who and what remains. The melancholic's refusal to relinquish their emotional bonds to who or what has been lost, Freud suggested, thwarts their ability to heal, to become whole again, and to confront their circumstances with the clarity needed to live a meaningful life in the wake of a loss.

Given his original understanding of mourning and melancholia, Freud's preference for the former and anxiety about the latter makes sense. His explanation for why mourning, rather than melancholia, opens onto healing should give us pause, however. What is it about severing our emotional attachments to the lost other that heals us? What is it about leaving the lost other behind that enables us to become "free" again? And why, in the end, should we accept that becoming "uninhibited" is superior to remaining mired in relations with the ones we have lost? Is there not something deeply unethical about a model of mourning that sees detachment, relinquishment, and disavowal as the end of grief? Rather than efficiently letting go or moving on, might we find healing in hanging on and staying with those who, and those which, we have lost?[78]

Within a few years of publishing "Mourning and Melancholia," Freud himself came to see the relationship between his titular terms in a new light, discovering they share more in common than he had at first presupposed. Whereas in the earlier text Freud characterized melancholic identification—the psychic process of setting the lost other up in oneself—as the principal impediment to recovering from a loss, in *The Ego and the Id* he conceded that even healthy mourning involves an identification. As it turns out, mourning is not simply a means of relinquishing the ties that bind the one to the other. In an extraordinary revision of his initial thesis, Freud concluded that "it is possible to suppose that the character of the ego is a precipitate of abandoned object-cathexes and that it contains the history of those object-choices."[79] The substance of who we are, in other words, contains the residual traces of all those other beings and ways of being we have loved and lost. In a sense, the lost other lives on within our memories, as well as in our words and our deeds. We do not abandon so much as we harbor who and what we have lost.[80] And it is in this propensity to safeguard the other from total loss that we discover the real ethical and political potential of grief and mourning.

To grieve is to acknowledge that a significant relationship has been dissolved, that a meaningful connection has been severed. Because we only grieve for those we care about, mourning reflects back to us—and to those who are privy to our grief—the relations within which we are entangled. "What grief displays," Judith Butler notes, "is the thrall in which our relations with others hold us."[81] The more we grieve, the more openly we acknowledge the ways our lives and our worlds are populated and enriched by others. This recognition is crucial for the work of earthly coexistence: all humans desperately need to own up to the profound fact that we are earthly beings enmeshed in earthly conditions with other earthly beings whose lives and deaths, whose existences, matter to our own. Ecological grief both confirms and affirms that we are not, finally, alone.

If ecological grief sheds light on the relational quality of earthly coexistence, it also illuminates the differential ways in which specific animals and plants, species, ecosystems, land communities, and earth systems matter to us. As Ashlee Cunsolo and Karen Landman explain, "who and what we do not mourn tells us as much about ourselves as who and what we do mourn."[82] We do not mourn all more-than-human beings and ways of being equally, just as we do not uniformly grieve all human lives.[83] Ongoing social movements coalescing around the affirmation, as well as the assertion, that "black lives matter" provide a potent and sobering example of the extent to which what Butler calls "grievability" is unequally distributed.[84] (The rejoinder that "all lives matter" both misunderstands and, paradoxically, proves the necessity of the original assertion.[85]) Just as publicly mourning the loss of black lives confers value on those lives by insisting that they mattered and continue to matter, so too does publicly mourning ecological losses imbue those more-than-human others with value.[86] Conversely, a failure to grieve a specific loss implies that there has been no loss—that nothing, or no one, significant has been lost.

While ecological grief retrospectively acknowledges the value of more-than-human beings and ways of being by demonstrating that a significant bond has been broken, it plays a role in our relationships with those who remain as well. This is so in at least two senses. First, we may experience "anticipatory grief." When we know that someone or something will be lost—that another being will die, that a species will become extinct, that a landscape will be wrecked, that an ecosystem will be fragmented—the grieving process often begins before the loss has occurred.[87] While awaiting the other's inevitable loss, we find ourselves confronted with options: Will we treat this threatened and vulnerable other's loss as a foregone conclusion? Or will we insist that no loss is unavoidable and, therefore, do what we can to stave off the other's death, extinction, disappearance, or destruction? In other words, acknowledging the other's potential or assured demise has its way of drawing us into an ethical realm, a space in which we are summoned to render judgments and make decisions about the form that earthly coexistence ought to assume. Perhaps this is why so much environmental writing reads like an elegy, not for what has already been lost but for what might yet be lost.[88] By projecting a future in which a loss may have come to pass, for instance, texts like Rachel Carson's *Silent Spring* ask readers to respond, in advance, to future ecological losses.[89] Only if we are able to foresee such losses do we become similarly capable of forestalling them. Grieving in an anticipatory mode, we realize that there is still time—time to care, time to intervene.

Second, our relationships with human and more-than-human others in the present are significantly shaped by whether we see these others as worthy of our grief and, thus, of our concern and our care. Not only do we mourn for those we care about, whose lives and ways of life we regard as valuable; we in turn care about and for those whose loss we know in advance we would grieve. Indeed, the presumption that we would grieve the other's loss furnishes the "reason" for supporting them in the present.[90] Accepting that we would mourn the other's loss places us in a position to do what we can in the present to delay, if not also to finally stave off, their loss. What Butler calls "grievability" manifests not in practices or rituals of mourning per se, but, rather, in the cultivation and sustenance of the social, cultural, economic, political, and ecological conditions that others need not only to survive but also to flourish.[91] Like anticipatory grief, the presupposition that another being or way of being would be mourned if it were lost inaugurates and orients relations of care based on the expectation of a loss—and a mourning—to come. One of the pressing challenges we face under current conditions of ecological loss and earthly transformation, then, is how to extend the reach of what the dominant culture considers to be grievable well beyond the human so as to initiate more loving, more care-full relationships with other animals and plants, species, places, landscapes, land communities, and earth systems.[92] Only when we conceive of such others as worthy of our grief are we likely to express the kinds of concern and enact the forms of care that earthly coexistence demands.

In *Being Together in Place*, geographers Soren Larsen and Jay Johnson provide a powerful example of how grief and a sense of grievability catalyzed a "decades-long fight" against the construction of a multi-lane highway—officially dubbed the South Lawrence Trafficway—through the Wakarusa Wetlands in northeastern Kansas, near the present-day city of Lawrence.[93] Home to several hundred species of plants and birds, as well as dozens each of reptiles, amphibians, fish, and mammals, the Wakarusa Wetlands has persisted as a biologically rich and culturally significant habitat for myriad Native American tribes for millennia. At the beginning of the nineteenth century, the Kanza, or Kaw, people—whose name became the base for the toponym Kansas—"ceded much of their land, including the Wakarusa Valley" through a series of treaties.[94] In 1882, a federal Indian boarding school was established in the Wakarusa Valley as a part of the settler government's attempts to "civilize" Native Americans and prepare them for assimilation into the dominant (that is, white European settler) culture. At the time of its establishment, the United States Industrial Indian Training School's 280-acre campus encompassed part of the wetlands that, though drained and converted into agricultural fields at the start of the twentieth century, gave the students, who had been wrested from their families and isolated from their cultures, a place to connect with one another and the land. A place where they could speak their languages and conduct their ceremonies.

A place where they could rendezvous with family. A place where they could carry on traditional relations with other animals and plants.[95]

By the time the South Lawrence Trafficway was proposed in 1985, the boarding school days had ended, and the school had been transformed into the Haskell Indian Junior College—a few years later it became Haskell Indian Nations University.[96] Over the years, Haskell's campus expanded and contracted, largely at the whim of the federal government; during the 1950s and 1960s, for instance, the Bureau of Indian Affairs transferred hundreds of wetland acres from Haskell to the University of Kansas and Baker University.[97] Despite these transfers, Haskell students and faculty carried on a connection with the wetlands. And it was this connection that fueled their response to the proposed trafficway. As Larsen and Johnson explain, their resistance to the trafficway emerged out of a sense of impending yet avoidable cultural and ecological loss:

> The realization that the [Southern Lawrence Trafficway] posed a direct and indirect threat to the life of the wetlands, the finitude, the death, and trauma of loss, actual and prospective, pushed people into a more vulnerable or susceptible state of awareness that prepared them for the call of place by connecting them to the other forms of life and modes of being that exist in that place.[98]

Catalyzed by a sense of grievability—by an inkling that they would grieve the Wakarusa Wetlands if they were lost—Haskell community members, as well as allies from Baker and the University of Kansas, campaigned against the trafficway for more than twenty-five years. While the trafficway was approved in 2012, the resistance to the project vividly demonstrates the ethical and political potential that ecological grief holds for the work of earthly coexistence. The fight for the Wakarusa Wetlands, as well as for the more-than-human beings and ways of being for whom it is home, emerged out of, and gained strength from, a collective sense of loss.

I began this section by briefly pondering the Latin root (*gravis*, "heavy") of the verb *to grieve*. Loss weighs on us, presses upon us. While grief can hold us in place, it can also move us by reminding us of all the ways we share our lives with others, human and more-than-human alike. Through our grief we acknowledge and affirm the constitutive nature of these connections, recognizing in our pain and our suffering how inextricably tied to others we remain. Because grief has a way of shining a light on those relationships that matter(ed) to us, it can serve as a pathway toward recovering and restoring relations. It can, as David Charles Gore eloquently puts it, become "a heart-sickness that heals."[99] And yet, our ready-to-hand understandings of grief all too often encourage us to think that the *telos* of grief is the restoration of a subject unbridled by connections to who or what has been lost. Yet, as I hope to have made clear here, we need not accept this perspective. Rather than liberating ourselves from the dead or disappeared, we can see grief as a way of sustaining connections with those whose loss we have survived. We can also see grief as an anticipatory stance, an orientation toward the vulnerability of the earth and our fellow travelers, that summons us to support and calls us to care. We could, in other words, work together to compose and maintain worlds in which we see the more-than-human as deserving of our grief now and in the future.

Grievable Rhetorics

"We grieve," the ecologist Aldo Leopold once remarked, "only for what we know."[100] Like Leopold, I believe that the more we know about the earth and our fellow travelers, the more likely we will be to regard them as grievable and, thus, to do what we can to stave off their death, destruction, or disappearance. As an ecologist and educator, Leopold prescribed time outdoors—in forests and fields; prairies and mountains; swamps and sloughs; along rivers, creeks, and lakes; among owls and oaks, white pines and wolves—as a remedy for what he saw in the 1940s, while finishing his *A Sand County Almanac*, as humanity's increasing alienation from the natural world.[101] Not everyone has experienced such alienation in the same way or to the same degree, of course. But for many people the sense of estrangement from the earth that Leopold detected so many decades ago has only grown stronger. Many among us barely know the names of our more-than-human cohabitants, scantly notice the changing seasons, and struggle to envision the earth as it once was and as it could one day be. Without these capacities, it is difficult to reckon with ecological losses past, present, and to come. In the rest of this book, I consider some of the ways we might cultivate and hone these capacities and, thus, some of the practices by which we might bring ourselves to grieve beyond the human.

Because I am a scholar of rhetoric, and since the practices I identify and think with throughout this book are rhetorical practices, a few words about rhetoric are in order here at the outset. The word "rhetoric" means different things to different people. It might mean ornamental, fluffy, misleading, or even deceptive speech to some. For others, it might mean the opposite of "action." How many times have we heard a politician castigate an opponent for spewing rhetoric yet failing to act? For still others, rhetoric might signal rousing oratory and compelling writing. For another group, rhetoric names the beating heart of deliberative, democratic societies. Though contemporary, each of these connotations has its roots in the intellectual and political culture of ancient Greece—where philosophers, such as Socrates, Plato, and Aristotle, debated rhetoric's risks and promises. Many of our negative ideas about rhetoric can be traced to two Platonic dialogues: *Gorgias* and *Phaedrus*.[102] In an especially suggestive scene of the *Gorgias*, for example, Socrates characterizes rhetoric as a "knack," as a kind of "cookery," and as a way of "pandering," all of which renders it "dishonourable undoubtedly" in Socrates' eyes.[103] By comparison, many of our more positive—or at least ambivalent—ideas about rhetoric arise from Aristotle's treatise on the topic. Whereas Plato greets rhetoric with suspicion, Aristotle approaches it with ambivalence: it can be used well or badly, for moral and immoral purposes, he explains, so we would do well to understand how it functions.[104] Carefully studying rhetoric, Aristotle demonstrated, sheds light on its inner workings and makes students better, more capable rhetors. All of these ideas remain with us today.

What makes rhetoric simultaneously loathed and loved, reviled and respected, is its association with influence. For better and for worse, rhetoric gets things done. Socrates, at a key moment in the *Phaedrus*, tells his young interlocutor that rhetoric is "a kind of leading of the soul by means of speech" in a wide variety of settings, from courts of law to political arenas to private settings.[105] If souls can be led, they can also be misled, as Socrates is quick to note. Tellingly, Socrates and Phaedrus then launch into a dialogue about deception that is meant to call attention to rhetoric's nefariousness. Meanwhile, in his treatise on the subject, Aristotle famously describes rhetoric as the "faculty of observing in any given case the available means of persuasion."[106] The ability to identify what would be persuasive in a particular context and for a specific audience, Aristotle argues, enables one to exercise power in public, political, and juridical settings. With our words we affect what others think, believe, feel, and value, and we influence what others do—whom they vote for, which organizations they donate their money to, what causes they give their time and energy to, and who and what they are willing to defend. In this sense, rhetoric is eminently practical. Of course, the rub is that all sorts of people—only some of whom have others' best interests in mind—put it to use. A double-edged sword, rhetoric can enrich as well deplete our lives. No matter how hard we try, we cannot rid rhetoric of its double nature as both poison and cure.[107]

The field of rhetoric as an academic discipline in the modern university emerged in the United States in the early twentieth century. No longer satisfied with studying classical literature, scholars and teachers interested in the role that communication plays in civil societies set out to reinvigorate the study and practice of public address—namely oratory.[108] "The initial study of rhetorical theory within twentieth-century communication studies," John Lucaites and Celeste Condit explain, "focused on the historical examination of classical and civic humanist models of persuasion and governance," which both imbued the fledgling field of study with "scholarly legitimacy" and provided a conceptual foundation on which to begin teaching the art of public speaking to undergraduate students—a tradition that continues to this day.[109] Drawing especially on Aristotle's analytic terminology, early rhetorical critics focused their attention upon single texts—usually speeches—in their original contexts, and worked to expose the operation of *ethos*, *pathos*, and *logos* within the text. This so-called "neo-Aristotelian" approach to the study, criticism, and practice of public communication dominated the field into the 1960s.[110] Since then, the discipline of rhetoric has undergone tremendous changes. While this is not the place to recount the history of those changes, suffice it to say that over the past several decades rhetorical scholars have vastly expanded both what and how they study.[111]

Among the transformations that have taken place within rhetorical studies since the 1960s, three are especially significant for the present study. First, the discipline has steadily become more amenable to explorations of environmental rhetoric. Beginning in the 1980s, a handful of rhetorical studies focused on environmental texts and crises appeared for the first time in mainstream scholarly venues like *The Quarterly Journal of Speech*. Since then, conferences, scholarly associations, and journals have been formed both to promote and to cultivate research and teaching on environmental rhetoric.[112] Second, the discipline has left behind the erroneous assumption that we can understand social, cultural, and political shifts by studying singular texts. Although there will always be good reasons to devote attention to exemplary texts, most rhetorical scholars now accept that rhetoric's influence is itself an "ecological" process, one in which a multitude of—often fragmented—texts interact and concatenate with one another to generate discursive worlds in which some feelings, beliefs, values, and actions seem more reasonable than others.[113] Finally, the discipline has become once again interested in the relationship among rhetoric, feelings, ethics, and politics. This is not a novel concern, of course; in the second book of his *Rhetoric*, for instance, Aristotle devotes considerable space to a detailed inquiry into the emotions.[114] And yet, recent years have also witnessed something of a renewed focus on the ways in which rhetoric modulates and even gives rise to our feelings.[115]

Building on these advances, I treat rhetoric as a suite of worldmaking practices—a material-symbolic toolkit for making things matter in particular ways; for bestowing value on certain dimensions of our shared worlds; for summoning our attention to specific issues; for generating affects, feelings, and moods; for shaping what we know and believe to be true; and for rendering some responses more probable than others.[116] This toolkit includes an assortment of implements—some verbal, some visual, some auditory, some bodily, some ambient—all of which can be put to work in the making and unmaking of shared worlds. I understand "worlds" as meaningful contexts—realms in which certain concepts, attitudes, behaviors, identities, objects, idioms, interpretations, and human and more-than-human others assume a kind of coherence and significance that they otherwise lack.[117] It is within worlds that we come to care about—in the neutral sense of being concerned with—some but not all others; it also within such worlds that we come to care for—in the not so neutral sense of taking care, supporting, and maintaining—some yet not all others. Rhetoric plays a crucial part in directing, though not determining in the strong sense, just who and what we see as worthy of our concern and our care.

Put differently, our feelings, thoughts, values, and actions are conditioned by forces extrinsic to and constitutive of the self. We are conditioned creatures, but we also take part in the construction of the conditions under which we dwell with others. "We are," as John Durham Peters memorably put it, "conditioned by conditions we condition."[118] Rhetorical practices help to shape those conditions and, therefore, participate in the crafting of various subject positions—myriad perspectives from which one comes to see, make sense of, feel about, and act into the world.[119] Through rhetoric, we generate the conditions under which particular attitudes, orientations, and commitments seem necessary or reasonable. We also create the conditions under which feelings are able to emerge. As Jeremy Engels explains, rhetorical practices "shape our emotions, and with our emotions, our worldview." "Only when we understand how to control our rhetoric," he adds, "can we develop some control over our emotions and hence how we see and interact with the world."[120] Knowing this, we are better equipped to comprehend what sets our earth emotions into motion and, therefore, how we might cultivate terranascient feelings like ecological grief.

The significance of this chapter's subtitle now comes into view. Whereas scholars, journalists, and activists routinely construe grief as a "natural" response to ecological and earthly losses, we need to be careful not to presume that ecological grief comes naturally.[121] Just because we should grieve for extinct species, imperiled land communities, or disrupted earth systems does not mean that we can or will. Just because an emotional response is appropriate, in other words, does not mean that we will be moved to feel it. Rather, our terranascient feelings are what we might call rhetorical achievements. It is not simply that we "have" the feeling first and then find ways to express it rhetorically, although we surely do this, too. More fundamentally, if and when we experience ecological grief, we can be certain that the rhetorical conditions that make this feeling possible have already been met. If and when we see another animal or plant or species, a place or landscape or ecosystem, as grievable, we can assume that the conditions which render this recognition possible have been organized and sustained. Instead of taking ecological grief for granted, then, or presupposing that people will eventually be moved of their own accord to feel the weight of the ecological and earthly losses that press upon us, I propose to consider the rhetorical practices by which a range of rhetors—authors, artists, and activists, chief among them—create the conditions under which we might be moved, and under which we might move others, to grieve beyond the human.[122]

Over the course of the next three chapters, I abandon the necessary abstractness of this opening gambit in order to consider in detail three rhetorical practices that, individually and collectively, set our ecological grief into motion. Each of these chapters opens with a vignette drawn from my own life; these prefatory narratives are designed to demonstrate the entanglement of specific rhetorical practices and the experience of ecological grief. My hope is that these vignettes will also inspire readers to reflect on their own experiences. Each chapter also includes a theoretical meditation; drawing from diverse sources, these conceptual forays provide a framework in which to consider how each rhetorical practice generates the conditions under which we are more likely to grieve ecological and earthly losses. The bulk of each chapter is devoted to the description and examination of concrete examples of each of the three rhetorical practices I investigate. I see my role as both curator and critic. I have tried, therefore, to bring before you cases that are exemplary in one way or another as well as to illuminate how they fit into the larger story I am trying to share about rhetoric's ability to compose worlds in which the more-than-human matters to us in ways that elicit our grief, our concern, and, ultimately, our care. They each do so, I contend, by orienting audiences toward ecological and earthly losses past, present, and to come.

In chapter 2, I consider the rhetorical practice of naming. Beginning with Jacques Derrida's claim that names foreshadow mourning, I argue that giving names to, as well as learning and uttering the names of, other animals and plants, species, landscapes, and even geological epochs exposes us to who and what we stand to lose.[123] Naming is one of the ways we make ourselves at home on earth; it is also, as I demonstrate, one of the ways we come to feel connected to what exceeds the human. It is precisely this sense of attachment to who and what could be lost that enables us to anticipate and, thus, to ward off the absolute loss of our fellow travelers and the earthly conditions we mutually depend upon. To substantiate these claims, I turn to an eclectic assortment of examples—from debates about what to call the last remaining member of a species in the scientific journal *Nature* to the proper names bestowed upon those solitary survivors; from artistic duo Ackroyd & Harvey's 2016 *Seeing Red.. Overdrawn* exhibition to a public scandal in 2015 over the disappearance of several names of plants and animals from the *Oxford Junior Dictionary*; from author Robert Macfarlane's *Landmarks* project to Barry Lopez's monumental *Home Ground*, both compendia of names for other earthly entities; and from activist and writer Bill McKibben's attempt to rename earth as "eaarth" to discussions about what name we ought to give to the current geological epoch. Together, these examples evoke the variety of naming practices and make plain the extent to which names capacitate us to anticipate ecological and earthly losses.

In chapter 3, I reflect on the rhetorical practice of archiving. Archives, typically understood as neutral repositories of the past, in fact furnish us with the rhetorical resources we need to narrate histories of the past and to imagine the future. In this chapter, I home in on phenology, the study of seasonal and cyclical natural rhythms, such as the spring arrival of American robins and the falling of oak leaves in autumn. As a professional and amateur pursuit, phenology involves its practitioners in a trio of activities: witnessing, documenting, and archiving. The detailed written chronicles of ecological events created by phenologists enable us to discern past transformations—including, for instance, the local extirpation of a species or a shift in the timing of seasons—and to detect ongoing perturbations. Although originally imagined as a means of revealing the inner workings of nature's stability, today phenological archives have become an essential source of information about the effects of climate change in specific places. In this chapter, I focus on the phenological research and writing of Aldo Leopold, whose handwritten notes documenting seasonal events in central Wisconsin during the 1930s and 1940s have, in conjunction with more recent records, shed light upon past ecological changes in the region caused by anthropogenic climate change. I also examine ongoing phenological research and archive-building undertaken by citizen scientists in a range of more and less organized contexts, from the USA National Phenology Network's national database of seasonal events to the Duluth Phenology Facebook page to the unwitting phenological archives accumulating on social media platforms like Twitter and Instagram. Phenological archives, I argue, not only enable us to grieve for past losses; they also prepare us for dwelling in a future characterized by still more losses—and, thus, more grief.

In chapter 4, I look at the rhetorical practice of making visible. One of rhetoric's enduring functions is to assist audiences in seeing what typically goes unseen, to pull back the metaphorical curtains and cast light onto usually invisible phenomena.[124] Since many ecological losses and earthly transformations occur slowly and subtly on the margins of human perception and attention, rhetorical practices which make losses past, present, and to come visible stimulate our capacity to imagine the lose-ability of what was, what is, and what could be. Four examples suggest the myriad ways in which visual rhetoric moves us to imagine and grieve the loss of the more-than-human and human worlds we inhabit: a simple wooden sign along a walking path marking the place where a five-hundred-year-old eastern hemlock once stood; a monument to the once abundant yet now extinct passenger pigeon in southern Wisconsin; massive hunks of Greenland's glacial ice placed in Copenhagen, Paris, and London by artist Olafur Eliasson; and a series of digital images depicting imagined civilizational futures in a world transformed by anthropogenic climate change. With these examples, I disclose some of the ways that visual rhetoric reveals the unseen and the unseeable. Each in their own way, these visualizations open our eyes and our imaginations to forms of ecological and cultural diminishment for which we are responsible and might come to care.

Finally, in the epilogue, I turn explicitly to the matter of care—what it means to care about and for the earth and our fellow travelers. In particular, I return to the question of how rhetoric opens onto a sense of ecological grief, which in turn furnishes us with the emotional energy we require to support and safeguard more-than-human others from loss. Oscillating between brief vignettes highlighting rhetorics of ecological grief and theoretical, personal, pedagogical, and political meditations, I argue that rhetoric provides a way of *caring for* earthly others by moving others to *care about* them. Through rhetoric, we take care of our only home and our earthly cohabitants.[125]

All told, this book lights up a tangle of intertwining pathways, each of which leads us in its own way to see other beings and ways of being as worthy of our grief. Rather than treating grief as an impediment to the work of earthly coexistence, this book places it at the center of this most persistent and pressing task. "If grief can be a doorway to love," Robin Wall Kimmerer teaches us, "then let us all weep for the world we are breaking apart so we can love it back to wholeness."[126] Amen. Let us also remember that our ability to grieve is not inevitable, that our mourning is conditioned, and that it is within our power to generate and sustain the rhetorical conditions that make our own and others' ecological grief more possible.

CHAPTER 2

Anticipating Loss

On Naming

Meditations on Naming

When my partner and I moved from Duluth, Minnesota, to State College, Pennsylvania, in the summer of 2020, we were delighted to discover that our rental home was just two blocks away from the southern terminal of a trail system that meanders through fields and forests and links our neighborhood, by way of Hartley Wood, to Pennsylvania State University's Arboretum. In a matter of minutes, we could leave the blacktop behind and enter the cradling embrace of mature stands of oak, maple, pine, and aspen, under which wildflowers bloom, groundhogs amble, deer feed, and birds dart from branch to branch. During our first summer in State College, at the end of many long days spent settling into our new house and trying to eke out a few hundred more words of the book you are now reading, evening strolls through the woods and arboretum became a near-daily reward—recompense for the days we devoted to unpacking boxes, building bookshelves, cleaning kitchen tiles, tending an unkempt yard, and, in my case, accumulating words on the page.

A year before we moved to central Pennsylvania, I acquired a pair of decent "bins," birding shorthand for binoculars, and an ornithological guidebook for Minnesota. Both sat at the ready on my desk beside a second-story window overlooking our front lawn, an old but sturdy silver maple (*Acer saccharinum*), and a file of columnar evergreen trees (likely a cultivar of *Thuja occidentalis*, or northern white cedar, though I never confirmed this). While I read and wrote, our cat, Judith, usually sat nearby, often snoozing by the window. Throughout the fall, winter, and spring before our move east, Judith informally accepted a new role. She became my BAS, my bird alert system. Better attuned to avian life than me, Judith would lunge at the window at even the slightest sight or sound of a bird. And so, when she leapt to get a better look, I followed suit. As often as not, I would spot what she had spotted and set about identifying the bird on the other side of the windowpane. The commonest birds I came to know by name quickly: American robins, ring-billed gulls, and crows, as well as common grackles, black-capped chickadees, brown-headed cowbirds, goldfinches, blue jays, and a few other birds were such routine visitors and so distinctive in appearance that, before long, I could identify and name them without my bird book. They had become familiars. Other birds eluded me, however, and required some effort to recognize and to name.

Though I am, at best, only a very casual and amateur birder, I derive a measure of pleasure from being able to identify and name the avian kin with whom I share a swath of land. Knowing their names—being able to say, "Do you see the cardinal in the lilac bush?" or "If you look closely, you can see a northern flicker working its way around the maple tree" or "I think that's a mourning dove"—makes me feel more connected, more familiar with the winged creatures who swoop in and out of my everyday life. Perhaps this is why so many people are attracted to birding, botany, and other kinds of naturalistic observation; the capacities we cultivate through these activities can make us feel more at home wherever we happen to be. I read in a popular birding book that, "We are . . . drawn to an abiding question: What is the name of that bird? A name is a tool for organizing our thoughts, for making sense of the world around us. Knowing the name of something makes it important. Giving a name to something immediately triggers a cavalcade of questions, of discovery, and of wonder."[1] Knowing the names of our more-than-human cohabitants, we feel less alienated from the earth and more interested in the multitude of others with whom we share the biosphere and the backyard.[2] While writing this paragraph, a skittish northern cardinal (*Cardinalis cardinalis*), a couple of confident grackles (*Quiscalus quiscula*), and American robins (*Turdus migratorius*) too numerous to count descended on the broken ground in my own backyard where, two days ago, my partner and I broadcast five pounds of oat and field pea seeds. That these seeds bring so many familiar birds to our yard is a gift from nowhere, and I greet our avian guests one by one, by name, as they arrive.

Hello, northern cardinal. Welcome, grackles. Greetings, robins.

Naming—giving names, learning names, and uttering names—prepares us for the work of ecological mourning.[3] By naming others, human and more-than-human alike, we anticipate their inevitable loss, their inexorable demise. Names announce the future disappearance of someone or something particular. Names gesture, always in advance, to the unraveling of specific beings and ways of being to come, to the death on the horizon, to the destruction that awaits us. And they do so because they disclose the possibility of speaking and writing of the other—and *for* the other—when they are gone, of sustaining a relationship with the lost other in the wake of their disappearance. Naming, simply put, draws us nearer to others in life and in death. The names we give, the names we learn, the names we utter: these are the vital, verbal conditions under which we come to know with whom and what we cohabit the earth and, thus, whom and what we stand to lose.

Yet, there is something ambivalent about naming. One always names the other. One presumes, therefore, the right—or the responsibility—to bestow a name on the other. One chooses, one invents, for the other their name. One decides, without consultation, *for* the other what they will be called. And so it is that, in naming, one subjects the other to a kind of containment by way of designation.[4] Perhaps this is why, at the very beginning of her best-selling book, *The Sixth Extinction: An Unnatural History*, Elizabeth Kolbert suggests that naming is partly responsible for the ongoing mass extinction event. "Beginnings," Kolbert writes, "are apt to be shadowy. So it is with this story, which starts with the emergence of a new species maybe two hundred thousand years ago. The species does not yet have a name—nothing does—but it has the capacity to name things."[5] The mere ability to name seems to imply danger. It is telling that Kolbert's story about anthropogenic mass extinction begins with the human capacity, and perhaps propensity, to name things. Rather than remaining neatly on the side of the symbolic, naming articulates with and props up a range of practices by which some humans carve up, categorize, organize, and manipulate the earth around us. We name others to bring them under our authority.[6]

Nowhere is the connection between naming and authority clearer than in the story of Genesis found in the Old Testament. Brimming with acts of creation and appellation, of generation and linguistic designation, Genesis is—as the title implies—about origins, about the formation of relations among God, people, and all the other living things on the earth. The very first act of appellation appears at Genesis 1:5, just after God has created lightness and darkness: "And God *called* the light Day, and the darkness he *called* Night." And so it goes with heaven, the earth, and the seas: all are made by God who then calls them by their names. But God also creates Adam and, in the second chapter of Genesis, endows him with the power and the responsibility to name all the other creatures which He has created: "And out of the ground the Lord God formed every beast of the field, and every fowl of the air; and brought *them* unto Adam to see what he would call them: and whatsoever Adam called every living creature, that *was* the name thereof" (Genesis 2:19). Amid this procession of making and naming, the author of Genesis writes, "And God blessed [Adam and Eve], and God said unto them, Be fruitful, and multiply, and replenish the earth, and subdue it; have dominion over the fish of the sea, over the fowl of the air, and over every living thing that moveth upon the earth" (Genesis 1:28). Naming and dominion, then, arrive together in the story of Genesis. Naming is the primary act of authority and, as such, founds the hierarchy that places God above humans and humans above our earthly cohabitants.

Some environmental historians treat Genesis as a kind of smoking gun, evidence of the relationship between naming and ecological destruction. In his now classic essay, "The Historical Roots of Our Ecological Crisis," for instance, historian Lynn White Jr. contends that the Judeo-Christian perspective on humanity's relation to the natural world is to blame for ongoing environmental catastrophes. Invoking Genesis, White remarks, "Man named all the animals, thus enabling his dominance over them."[7] Symbolic and material control converge, for White, in the act of naming. To name something is to get a hold on it, to bring it under one's power, to make it available for manipulation. To name another is to transform them into an object that can be scrutinized, identified, categorized, known, managed, used, and abused. We can order and organize what we can name. Simply put, naming is taming. "That which will not be named," Neil Evernden contends in *The Social Creation of Nature*, "cannot be controlled."[8] If it is true that naming and controlling are so closely aligned as to be almost indistinguishable, then science is as much to blame for our ecological crisis as is the Judeo-Christian perspective. Western science is obsessed with "discovering" and naming species of animals, plants, and fungi. As the biologist Michael Ohl notes in his *The Art of Naming*, "Approximately 1.5 million species have been identified, catalogued, and named, but an additional three, five, ten, or even 100 million are waiting to be found. And named."[9] Framed in these terms, naming other species is one practice in a broader effort to uncover and record what exists—for the sake of knowledge, to be sure, but for purposes that may well run counter to the flourishing of other species, too.

If naming other species of animals and plants can function as a means of dominating beings and ways of being, naming and renaming places can serve as a means of controlling the land itself. In the present-day United States, settler-colonialism's appropriative logic is wrapped up with acts of appellation. As George Stewart writes in *Names on the Land*, when European explorers descended on North America, Native American toponyms already "lay thickly over the land."[10] For many Native Americans, toponyms do more than reference; they also carry histories, constitute ways of knowing, and sustain vital connections among people and places. The Western Apache, for example, use an astonishing array of toponyms to "summon forth an enormous range of mental and emotional associations—associations of time and space, of persons and social activities, of oneself and stages in one's life."[11] Similarly, for the Tlingit, toponyms can evoke "human tasks, events, emotions, and other mental associations," thereby imbuing particular places with significance and ensuring that cultural and environmental histories and knowledge are passed from one generation to the next.[12] Consider two names for the same stream in Juneau, Alaska: the English name that appears on most maps is Gold Creek; the Tlingit name is *Dzántik'i Héeni* (Flounders at the Base of the Creek). Whereas the English name, concocted by miners, identifies gold as the stream's defining feature, the Tlingit name characterizes the creek as a place where people and flounders encounter one another. Whereas "Gold Creek" frames the stream as a place to be exploited for financial gain, in other words, "*Dzántik'i Héeni*" calls our attention to the subsistence relations that have long been at play in this place.[13] One name is a summons to ruination, the other a guide to ongoingness. The differences matter.

"There is power in naming and renaming," writes Winona LaDuke (Ojibwe). "That power is widely abused in the United States."[14] Nominal abuse is pervasive and takes many forms. Through transliteration, for example, settlers have mangled many Native American toponyms; carelessly translated toponyms distort the meanings embedded in the names and obscure the cultural and ecological wisdom they were created to evoke.[15] The violence cuts still deeper, though: the US government has also actively attempted to deprive Indigenous peoples of their languages and, with them, the place names their ancestors bestowed on the land. Some of these attempts have been direct; the US Office of Indian Affairs' boarding school program, for example, explicitly prohibited young people from speaking their native languages with one another.[16] More generally though less directly, settler colonial violence aimed at annihilating peoples leaves fewer speakers to preserve their languages, leading to a sad state of affairs in which, LaDuke notes, "most North American Indigenous languages are expected to be extinct by the year 2050."[17] When a living language, especially one that is primarily oral, becomes extinct an entire way of knowing and seeing the world vanishes with it. As one Tlingit elder explained: "'Lyee sakoowoo saawx' ch'a tleix ee jeedax goox la hash ee koosteeyi.' (If you don't know the names, your [Tlingit] way of life will drift away forever.)"[18] As this Tlingit elder's comment suggests, the power to name and rename is intricately entwined with the power to decide which ways of life will likely survive and which will likely disappear.

The loss of Indigenous toponyms and other names is especially worrisome for two reasons. First, names are one of the primary means by which many Native peoples establish and maintain connections to particular places and more-than-human others. As Potawatomi botanist Robin Wall Kimmerer notes, "Names are the ways we humans build relationship, not only with each other but with the living world. I'm trying to imagine what it would be like to go through life not knowing the names of the plants and animals around you." This thought experiment leads Kimmerer to conclude that, "I think it would be a little scary and disorienting—like being lost in a foreign city where you can't read the street signs."[19] This points to a second and related reason that losing Native appellations should worry us. When Indigenous names are lost, the traditional ecological knowledges they bear and express are likewise imperiled. When settlers rename what has already been named, they constrain our collective ability to respect and learn from not just the names themselves but also the wisdom embedded within them. "In their place-making," Lauret Savoy explains, "these newcomers not only set out to possess territory on the ground. They also lay claim to territory of the mind and memory, to the future and the past."[20] In other words, the loss of Indigenous place and other names depletes our collective storehouse of cultural resources for engaging in the work of earthly coexistence and diminishes our ability to flourish alongside and with others.

There is much more that I could say about the potential violence of naming, about the links between designation and domination, about the ways in which naming others is an exercise in political, ethical, and historical generations and erasures. I am inclined to see naming as a form of what the environmental philosopher Thom van Dooren calls "violent care," an act that inevitably muddies the boundaries between harming and helping.[21] Surely naming can alienate us from other beings and ways of being, just as renaming can estrange us from sedimented histories of place and landscape. There is always a certain violence at the core of naming, for everything that has a name could have been named otherwise. To choose one name is to refuse so many others, to highlight some qualities at the expense of others. But, as Indigenous and non-Indigenous thinkers alike demonstrate, naming can also draw us nearer to, and so bring us into more intimate relationships with, human and more-than-human others. To give, to learn, to utter a name—these practices are some of the ways we express care and concern.

Moreover, as I show in the following pages, naming can set into motion and sustain ecological grief. Naming anticipates losses to come, thereby enabling us to anticipate who and what are at stake. When we are able to name the beings and ways of being, the places and the habitats, the ecosystems and the earth systems, that our actions threaten, so too are we able to safeguard the precarious from these harms. If I cannot name the other, how can I struggle with and for them? How can I speak on behalf of someone or something whose name I do not know and, thus, cannot utter? With these questions in mind, let us consider how naming prepares us for ecological grief and, thus, how it prepares us for the difficult yet necessary work of earthly coexistence.

Several years ago, while reading *The Animal That Therefore I Am* for the first time, I was struck, like many readers, by Jacques Derrida's description of his encounter with a cat in his bathroom. Naked, Derrida stands before this other for whom nudity cannot be a concept, for whom there can be no meaningful distinction between being naked and being clothed. Despite this, in front of this cat Derrida feels ashamed of his body, especially his sex. Face to face with this very real cat with whom he shares his home, Derrida remarks:

> Nothing can ever rob me of the certainty that what we have here is an existence that refuses to be conceptualized [*rebelle à tout concept*]. And a mortal existence, for from the moment that [this cat] has a name, its name survives it. It signs its potential disappearance.[22]

This cat who has a name—who has, perhaps, several names at once—looks back at Derrida, regards him. This cat who has a name is a singular being, a wholly other Other with whom Derrida can relate but which he cannot "conceptualize," which he cannot transform into an idea. No singularity lends itself to generalization. This cat who has a name is, Derrida tells us, a "mortal existence." Interestingly, this cat's mortality—its status as the kind of being who can and will die—is disclosed by its name, by the fact that it bears a name. And this is so, we learn, because the "name survives it" and "signs its potential disappearance." The name outlives the living; it anticipates the loss of that which bears it.

Names announce a loss to come, an unpredictable yet certain disappearance. Names declare, in advance of any actual loss, that the being or way of being which bears this name will, someday, for some reason, wither away. Thus, every name is saturated with pathos, with a certain sadness. Every name, Derrida writes, is "*a foreshadowing of mourning*." And this is so, he explains, because each

> case of naming involves announcing a death to come in the surviving of a ghost, the longevity of a name that survives whoever carries that name. Whoever receives a name feels mortal or dying, precisely because the name seeks to save him, to call him and thus assure his survival. Being called,

> hearing oneself being named, [and] receiving a name for the first time involves something like the knowledge of being mortal and even the feeling that one is dying.[23]

Ghostly, spectral figures, names linger after the death, destruction, or disappearance of the other. As Derrida saw it, naming is a means of conjuring ghosts, of ensuring that the dead never fully leave us. (They also enable *us*, survivors who will one day be survived, to tarry with the dead, to speak of and with who or what has already been lost.) From the beginning, the name serves as the durable trace of the who or the what to which it is attached. Thus do we place the name upon the headstone; each body must be identified, by name, so that survivors may return, again and again, to *this* one in particular. Names go on referring and signifying in the other's absence. Indeed, the name always signifies the other's (potential) absence.[24]

Which is to say: the name's work is never done. Herein lies naming's special power to "save" the other. In their obstinate persistence—even if lost or forgotten or covered over, the name can always be recovered, found again, spoken anew—names prevent a total loss. Even when I die, my name—along with all its metonymic associations—will persist. This is as true for proper names as it is for the names we give to entire species, to landscapes, to places, and even geological epochs, as we will see at the end of this chapter. More than a hundred years after the last one died in captivity at the Cincinnati Zoo, for instance, we can and do still speak of the passenger pigeon (*Ectopistes migratorius*). In every case, the name has already reached ahead to a time after the loss, to a time when this particular being or way of being will *be* no longer. There will come a day when we speak of the Greenland ice sheet in the past tense. The name will make this possible. A kind of insurance, names ensure that the loss of matter is not accompanied by the loss of memory. For survivors, names tether us to what has already been lost; the name is a linguistic rope onto which we can cling. Long after the other has died or disappeared or been destroyed, its name endures. And because it persists, we are able to hang onto that which has passed, or been pushed, out of existence. Simply invoking the other's name is sometimes enough to conjure them, to make them present in their absence. Names sustain our bonds with the lost, the extinct, the disappeared; they save these ways of being from absolute annihilation. It is thus that a deeply preservative tendency underwrites the practice of naming.

Naming not only enables us to hang onto who or what has already been lost, though. In an important yet often overlooked sense, naming also orients us toward the future loss of what remains, to the impending disappearance of who and what are still with us. In the name we are able to discern the other's mortality and, thus, their lose-ability. The presence of the name always gestures toward the real and potential absence of its referent. Whether we know the name "snowy owl" (*Bubo scandiacus*), whether we can connect that name to a fleshly being, the snowy owl's existence remains precarious. When we know the name, however, and when we utter it, we open ourselves to a way of being whose persistence is by no means guaranteed. Simply put, if we cannot—or if we refuse to—name those lives and ways of life that are threatened, we also cannot work to protect them from harm. In the absence of the appellation, it becomes difficult, if not impossible, to speak or work on the other's behalf. And so, there is a poignancy inherent in the act of naming. That this being or way of being bears a name at all implies that someone found it worthy, deserving, or in need of a name. The presence of a name, in other words, implies value. Someone took the time to craft this name, *Bubo scandiacus*. That we come to know, and perhaps to utter, this appellation affirms that this species is valuable, worthy of our words, if not also our deeds and our grief.

Naming plays an essential role in Derrida's preemptive politics of mourning. In the wake of loss, we surely grieve. But we also grieve in the face of losses to come. Reflecting on friendship, Derrida often remarked that grief insinuates itself into every loving relation from the very beginning. Indeed, as Joan Kirkby contends, Derrida's great insight was that "all our relationships are *from the beginning* tinged with mourning."[25] This is so because every relation is marked by the anticipation of a singular loss. On this point, we can do no better than to listen to Derrida himself, who once wrote:

> To have a friend, to look at him, to follow him with your eyes, to admire him in friendship, is to know in a more intense way, already injured, always insistent, and more and more unforgettable, that one of the two of you will inevitably see the other die. One of us, each says to himself, the day will come when one of the two of us will see himself no longer seeing the other and so will carry the other within him a while longer, his eyes following without seeing, the world suspended by some unique tear, each time unique, through which everything from then on, through which the world itself—and this day will come—will come to be reflected quivering, reflecting disappearance itself: the world, the whole world, the world itself, for death takes from us not only some particular life within the world, some moment that belongs to us, but, each time, without limit, someone through whom the world, and first of all our own world, will have opened up in a both finite and infinite—mortally infinite—way.[26]

Every relationship is marked, from the very beginning, by the possibility of loss. The name signifies who or what will one day disappear, the being or way of being whose loss we will have to endure, and whose loss we commit to enduring from the moment we learn their name. The name preempts a loss to come, setting our grief into motion before the other has disappeared. We await death, loss, disappearance, destruction: we wait, that is, *already in mourning*, for a time to come when we will surely grieve.

Thus far, I have been making my argument in positive terms: here is what naming makes possible; here is how naming prepares us to grieve. There is another, more negative way to make the point, however. When we do not know the names of those others that find themselves threatened by erasure, we have a hard time apprehending, much less mourning, their potential loss. How can we possibly grieve what we cannot name and, thus, what we can neither identify nor discern, what we cannot distinguish from all else? In her work on grief and grievability, Judith Butler repeatedly affirms the significance of the name for both setting into motion and sustaining our mourning. And, crucially, Butler accentuates the fact that, without a name, we struggle to grieve. "I am as much constituted by those I do grieve for," she writes, "as by those whose deaths I disavow, whose nameless and faceless deaths form the melancholic background for my social world."[27] These nameless deaths hardly register; indeed, they may not register at all. There are surely so many others whose names we do not know and whose absence we do not grieve, whose names we never hear or learn or utter and, thus, whose loss registers as no loss at all. In times of war, Butler notes, we rarely hear the names of those who have died on "the other side." She asks, "Will we feel compelled to learn how to say these names and to remember them?"[28] And, if we refuse to do so, do we thereby relegate these nameless others to the ungrievable? When there is "no image, no name, no narrative," Butler worries, it often seems "that there never was a life, and there never was a death."[29]

If our ability to see others as grievable—as worthy of our grief and, by extension, our concern and our care—depends in part on whether we know and utter the other's name, a certain political and ethical path reveals itself to us. "Language deficit leads to attention deficit," writes Robert Macfarlane in *Landmarks*.[30] We grieve who and what we can name, and we tend (mentally and materially) to who and what we would grieve. The earth needs our attention. And so, we need to cultivate our capacities for naming what is at stake, what our being threatens to diminish or erase, what our collective action might spare from ruination. To be sure, names are only one of the rhetorical conditions under which we come to see the more-than-human as grievable. But names are essential for grief. They direct our attention to what is with us and, therefore, to what we can lose. It is a matter of discernment. It is the difference between being able to say that "some birds" visited your yard and being able to identify, by name, the species that flew into your lawn and, therefore, your life. The difference matters, for we can only grieve and struggle for such *particular* beings and ways of being on earth.

Naming What We Might Lose

When I think about the extinction of the passenger pigeon—once the most abundant species of avifauna in North America, if not also on earth, with a population reportedly stretching past five billion—I cannot help but recall that the very last living specimen of *Ectopistes migratorius* with whom humans shared the earth died, alone in an aviary at the Cincinnati Zoo, on September 1, 1914. Four years prior, the Cincinnati Zoo's second-to-last specimen, a male named George, died, his body apparently thrown out with the trash. From 1910 to 1914, the last known passenger pigeon lived alone.[31] No one with whom to mate and, thus, no offspring to leave behind. No companion with whom to pass the days and nights. No kin left and no familiar cooing in the final moments. It is a sad story. To have one's death coincide with the extinction of one's species, to be the last one of your kind to leave the earth—such is the fate, and the burden, of "endlings."

The word "endling" emerged in the mid-1990s when readers of *Nature* discussed, in the journal's Correspondence section, what we should call the last survivor in a family line or of a species. Robert Webster and Bruce Erickson of the Christian City Convalescent Center in Union City, Georgia, considered other terms—"lastoline," "endler," "endmost," and others had occurred to them—but they settled on "endling" to describe "the last person, animal, or other species of his/her/its lineage."[32] Their proposal inaugurated a small debate in *Nature*. Other readers chimed in with proposals of their own: "terminarch," "relict," and "Mohican," for instance, were all suggested.[33] The last of these, "Mohican," builds on and reinforces the "vanishing Indian trope," which describes Native Americans in the past tense and suggests that the Native inhabitants of present-day United States have all "vanished."[34] The image of the vanishing Indian is empirically false; several million people living in the United States today belong to one of the more than five hundred Indigenous nations. But the image is also perniciously misleading in another sense; it is misleading insofar as it obscures the fact that the settler government, often in conjunction with individual vigilantes, has actively attempted to disappear Native peoples. Fortunately, "Mohican" did not catch on.

Instead of "Mohican" or "terminarch" or "relict," Webster and Erickson's offering prevailed. In the more than two decades since they invented the word, the historian Dolly Jørgensen notes, "endling" has catalyzed historical, artistic, and political practices, from museum exhibits to musical compositions to literary projects. The word's power, Jørgensen contends, lies in its capacity to "make the narrative [of extinction] personal while retaining the universality of extinction—when this individual is gone, the whole species is no more."[35] When there is only one left, it becomes impossible to disentangle the individual from the species. To lose one is to lose the other. As Bruce Erickson remarked years after he helped introduce "endling" to the world, "We don't name the things we choose to ignore. So, somehow, naming it gives it a value that wasn't there before. If that's what the word is doing, I'm really proud to have been a part of it."[36]

Which is what happened on September 1, 1914, when the last surviving passenger pigeon died in her aviary—a term that obscures the fact that she, like all zoo animals, lived in a cage—at the Cincinnati Zoo. She, alongside the handful of known endlings with whom she shares this sad title, was given a proper name: Martha. The names Martha and George, it is said, were chosen in homage to the Washingtons. In environmental writing and anti-extinction literature, though, she is often referred to as "Martha the Passenger Pigeon." Of course, the "the" in this title must do double duty; on the one hand, it denotes the species to which Martha belonged and, on the other, it calls attention to the conflation of individual and species that characterizes the life of every endling. For a long, few years, Martha was literally *the* (only) passenger pigeon. Incas the Carolina Parakeet died in the same aviary as Martha on February 21, 1918. Benjamin the Tasmanian Tiger perished on September 7, 1936. Lonesome George, the last surviving Pinta Island tortoise, died in captivity on June 24, 2012. As with Martha, the deaths of Incas, Benjamin, and Lonesome George marked the end of an individual life as well as the life of a species. "They are all dead now and can never be replaced," Mike Shanahan observes, "but at least they got names." "Their names remind us," Shanahan adds, "that we have epic tales to tell of the decline and fall of entire species."[37]

In the proper names given to endlings like Martha, we find ourselves exposed, then, to the potential disappearance of both a singular being and an entire way of being on earth. Collapsing the division between individual and species, the proper names we bestow upon endlings disclose both at once. "Martha" metonymically signifies just this being *and* all the others like her that have already come and gone. The proper name anticipates loss from the moment it is given, from the moment it is uttered. Whoever or whatever bears such a name can be lost and, thus, can be mourned. In *The Politics of Friendship*, Derrida remarks that the proper name "weeps death before death."[38] But in the case of endlings, proper names gesture toward species-level extinction as well, toward the absolute lose-ability not just of particular plants or animals but of species as a whole. The pathos attached to the singular life of an endling, the poignancy of their lives and especially their deaths, spills over to all the beings like *this* being whose death is, as it were, the final nail in the proverbial coffin. Temporalizing and personalizing extinction in this way, the proper names given to endlings render the finality of extinction in terms we can understand, on a scale we can comprehend. As Eric Freedman remarks in the *Earth Island Journal*,

> I have trouble wrapping my mind around the mega-threats and the huge numbers of species on the brink. The figures are simply too large. They make the situation seem inevitable, uncontrollable, and impersonal. There is none of the stark clarity of zero—the reminder that extinction is forever.[39]

We are accustomed to thinking about death as something that happens to individual people with names. Many of us struggle to comprehend the scale of extinction, which is about more than the loss of any specific person or life. Extinction is often an anonymous—which is to say, a nameless—affair. Thus do proper names such as "Martha" and "Incas" and "Lonesome George" bridge the gap between the "mega-threats and the huge numbers" and "the stark clarity of zero."

Bestowing proper names on threatened animals and plants does not ensure that we will protect them from extinction. Martha, Incas, Benjamin, and the rest died; their species became extinct. The name alone can save neither a species nor an individual from loss. By the time an endling is named, after all, it is typically far too late, their fate having already been sealed. And, of course, a single specimen—even a few specimens—does not a species make. Understood as a distinct way of life, a species requires more than a few specimens to flourish. So, as the environmental philosopher Thom van Dooren cautions, there is a risk in focusing on endlings because "the immensity and significance of extinction cannot be captured within these singular events."[40] Even as we grieve for endlings, he contends, "the deaths of these last individuals must be understood as singular losses in the midst of the tangled and ongoing patterns of loss that an extinction *is*."[41] The death of an endling is perhaps inherently poignant. It is perhaps easier to feel sad about, and to grieve, the death of a charismatic, lonely bird named Martha than it is to grieve the loss of billions of other "nameless and faceless" passenger pigeons. Yet, if we attend only to the last living member of a species, we miss the species itself. We reduce the complex way of life that a species is to a singular individual. We confuse death, which is inevitable, with extinction, which is often the result of overlapping disturbances and historical violence.

While van Dooren is right to caution us against focusing too acutely on individual deaths in a time of mass extinction and against the forms of sentimentalism that may arise when we look at extinction through the more familiar lens of organismic death, the proper names by which we often come to know the last of a species nevertheless play an important role in our capacity to grieve, and thus to care, for those species who find themselves on the edge of extinction. Many animals threatened with loss now receive proper names. For instance, Najin and Fatu, the last of the northern white rhinoceroses (*Ceratotherium cottoni*), have received attention from journalists, scientists, and environmentalists alike. Especially since the last surviving male, Sudan, died in 2018, the fate of the northern white rhinoceros seemed secure. Extinction seemed certain. Either Najin or Fatu will live to see the other die; one of the two of them will join the growing roster of endlings, a partial list of whom can be found on a Wikipedia page dedicated to this group. When we utter their names, then, we announce the individual and collective losses to come. The name says, in advance, that these beings, and the ways of being they metonymically represent, will perish.

Of course, many people hope to stave off those losses, to prevent the extinction that every endling's precarious existence would seem to foreshadow. And, while unlikely, it is possible that the northern white rhino could be pulled back from the edge of extinction. "The only way to save the northern white," Ed Yong writes in *The Atlantic*, "is to now artificially inseminate their eggs with stored sperm from Sudan and other males, and [to] implant the resulting eggs into females of the closely related southern white rhino."[42] So it is that Najin or Fatu's offspring, carried by a member of a related species, might be born into a world they will surely leave behind. If that ever happens, it is likely that these young northern white rhinoceroses will be given names, too—names that will remind us of their singularity, their precarity, their lose-ability; names which will invite us to see these new beings as worthy of our grief.[43]

If the proper names ascribed to endlings bring the immensity of extinction down to scale, articulating species loss through individual death, a 2016 work by UK-based artistic duo Ackroyd & Harvey, titled *Seeing Red.. Overdrawn*, intentionally emphasizes the enormity of the losses to come by deploying names in an altogether different way. Displayed during the pair's *Conflicted Seeds* show at the University of Cambridge's David Attenborough Building, *Seeing Red.. Overdrawn* suggests the power of names to draw our attention to species that will surely disappear without an active effort to preserve them. Upon a twenty-two-foot-wide by nine-foot-tall white canvas, the artists presented the Latin and common names (where available—not all species have been given a common name) of the 4,734 species of plants, animals, and fungi that were then categorized as "critically endangered" on the International Union of Concerned Scientists' Red List.[44] Each of the Latin names was printed in faint red Caslon Italic, a decorative serif font, while the common names were printed in gray Gills Sans Uppercase, a font characterized by straight, unadorned lines. Following Charles Darwin's lead, the species' names were divided from one another by two consecutive periods. Barely perceptible from a distance, the names of the 4,734 critically endangered species became legible only as spectators drew nearer to "the writing on the wall." Approaching the work, these spectators were invited to become participants, to draw over—in the intense hues of red and black permanent markers—each of the names and, thus, to bring each of the fading species into sharp relief.[45]

Distance and proximity are key to the rhetorical force of *Seeing Red.. Overdrawn*. One can take in the immensity of the extinction crisis from afar by standing back from the wall of names and surveying the extent to which our earthly cohabitants find themselves at risk of erasure. From a distance, the hints of red and black barely register as names at all. The faded text lends texture to the wall, but the names do not stand out. As if disappearing before our eyes, the names of these endangered species seem to be themselves endangered, at risk of vanishing. And the further one gets from the wall, the harder it becomes to make out the catastrophe unfolding before us. But this is more than an optical illusion or visual gag. Rather, *Seeing Red.. Overdrawn* forcefully suggests a critical link between the 4,734 names on the wall and the ways of being on earth they represent: if one fades, so too might the other. Reflecting on Ackroyd & Harvey's work, Akiko Busch wonders: "What becomes extinct first—the object or the word? Is it possible for one to rescue the other?"[46] By saving one, might we also save the other? By learning and uttering the names of the threatened, can we do more than symbolically save our endangered earthly cohabitants from tumbling over memory's edge? By saving the names from disappearance, can we also safeguard the ways of being they represent from being lost?

Seeing Red.. Overdrawn suggests that safeguarding critically endangered species from extinction begins with rescuing their names from obscurity. As spectators move closer to the wall, the undifferentiated texture of faint red and black traces transforms into a litany of distinct names. The nearer spectators stand to the wall, the fewer names they see at once. Yet, as the quantity of scientific and common names contracts, the intensity of each specific name increases. The closer one gets to the wall, the more volume of information gives way to richness of data. Drawing nearer, one begins to discern not just the immensity but also the intimacy of who and what are at stake. For example:

> *Hyloxalus ruizi* RUIZ'S ROCKET FROG .. *Icterus oberi* MONTSERRAT ORIOLE .. *Idiostatus middlekauffi* MIDDLEKAUF'S SHIELDBACK KATYDID .. *Lepidopyga lilliae* SAPPHIRE-BELLIED HUMMINGBIRD .. *Leucogeranus leucogeranus* SIBERIAN CRANE, SIBERIAN WHITE CRANE, SNOW CRANE .. *Melomys rubicola* BRAMBLE CAY MELOMYS, BRAMBLE CAY MOSAIC-TAILED RAT

As even this brief excerpt demonstrates, the names beg to be studied and read, each one a small, suggestive prose poem that opens spectators to a way of being beyond the human. But the names also beg to be accentuated, distinguished one from another and from the white background that bears them all. "One member of the public at a time is invited to make their mark by over-writing with indelible pen the name of one of the species," the artists explain on their web site, "drawing attention to it, and bringing it to visibility."[47] With red and black markers, participants carefully trace over the printed words, lifting each name up off the background and letting it be seen more clearly. This over-drawing is a means of rendering each name and each species legible, of highlighting the thousands of ways of being on earth that are at risk of disappearing.

As each intimate act of over-drawing accumulates, the once-muted wall of names becomes more and more saturated with both color and significance. One after another, the names intensify. The transformation of the wall is poignant for several reasons. First, each of the names has had to be individually and carefully re-inscribed. Some of the same human beings whose existence puts these 4,734 endangered species at risk have cautiously traced the faint lines in permanent ink. In a small way, this act establishes a personal connection: participants learn a new name, a new species, and thus a new story of potential loss. The care that one must exercise while tracing over the printed letters mimics, again in a small way, the care and attention to detail that critically endangered species deserve.

Second, the process of re-inscribing one of the names shrinks and expands one's sense of the extinction crisis. On the one hand, in order to write over the printed ink, one must focus one's attention narrowly on a single name, on a single letter at a time. Yet, on the other hand, because these are scientific and common names instead of proper names, to engage these appellations is to encounter not a specific being but an entire way of being on earth. Each of the names on the wall metonymically gestures toward not only an array of individual animals, plants, and fungi but also (without elaborating) the distinctive ways that these beings inhabit the earth.

Third, and crucially, as each name is traced over, both a potential disappearance and a potential recovery are announced. Depending on what we do—and do not do—the writing on this wall will serve either as a commemorative artifact that reminds us of all we lost or as a banner reminding us of all we saved. To confront the 4,734 names inscribed on this wall is to encounter both possibilities. Each name poses the questions: Will you grieve us if we are lost? Will your sense of impending grief compel you to act on our behalf so that we, too, might survive?

Seeing Red.. Overdrawn derives much of its rhetorical force from the way that the wall merges and contrasts individual names with the accumulation of appellations. The Vietnam Veterans Memorial in Washington, DC, works analogously. Designed by then twenty-one-year-old architect Maya Lin and finished in 1993, the Vietnam Veterans Memorial consists of two approximately 250-foot-long black granite triangular walls depressed into the ground near the Lincoln and Washington Memorials. The V-shaped memorial reaches its ten-foot-high apex where the two walls converge; as the walls extend outward from this climax, though, they descend to a height of just eight inches. From afar, the memorial's 140 panels appear textured but text-less. As one approaches the memorial, however, one comes to see that each panel is etched with the names of those who died in the Vietnam War. "When one realizes that they are not texture but the names of the dead inscribed along the entire stretch of the wall," writes Robert Pogue Harrison in *The Dominion of the Dead*, "one is struck by the sheer profusion that makes up this class of veterans who met their death in Vietnam." In a double gesture, the more than 58,000 names etched into the wall point both to the accumulation of lives lost in the war and to the unresolvable singularity of each specific death. "When we speak of 59,000 dead," Harrison continues, "that is a crushing figure even in its abstraction. But allow their names to occupy space and suddenly the 'so many' undone by the war find an immoderate measure for their excess."[48] While descendants of veterans search for their lost family member, they inevitably encounter the names of hundreds, perhaps thousands, of other people who died in the war. Each name points to an irreplaceable, irretrievable some*one*; together, the names—"seemingly infinite in number," as Lin described them—resonate to indicate the sheer volume of loss.[49]

There is, however, a crucial difference between *Seeing Red.. Overdrawn* and the Vietnam Veterans Memorial (and other commemorative sites, such as the US Holocaust Memorial Museum, that display the names of the dead on the wall): the names on the wall at the University of Cambridge are not of the dead but of the *dying*, not of the already gone but of those ways of being we *might yet lose*.[50] They call us not to grieve extinct species but to see these species as worthy of our grief. They invite us, as Derrida once put it, to "weep death before death."[51] *Seeing Red.. Overdrawn* wagers that an encounter with the possible losses to come will inspire us to care for these ways of being so that they will not be lost. Whereas the Vietnam Veterans Memorial commemorates the more than 58,000 Americans who died in the Vietnam War, preserving their names but not their lives, the 4,734 names on Ackroyd & Harvey's wall seek to safeguard both the name and the way of being, both the appellation and the way of life for which it stands. Indeed, *Seeing Red.. Overdrawn* speaks to us in the subjunctive voice, asking "what if?" *What if* we learned these names? *What if* we uttered these names in public? *What if* we came to recognize the beings and ways of being these names gesture toward as worthy of our grief and, thus, of our care? *What if* we labored to prevent this wall of 4,734 names from becoming, like the Vietnam Veterans Memorial, a work of commemoration? *What if* we stopped these names and all they stand for from disappearing?

Names—like the beings, ways of being, and places they stand for—are precarious. Names can be misplaced, forgotten, replaced, or erased, as the earlier discussion of Native place names demonstrated. In the world of words, names are comparatively unstable: what we call someone or something today may not be what we call them tomorrow or in another fifty years—if we call them at all. Names are not forever on the tips of our tongues.

In a rather unusual way, the precariousness of names, and especially the names of more-than-human ways of being on earth, surged into public consciousness in 2015. Nearly a decade earlier, in 2007, the Oxford University Press released a new edition of its *Oxford Junior Dictionary*, a compendium of roughly five thousand words its editors deemed useful or of importance to seven-year-olds. The small hands of its primary readers require that the *Junior Dictionary* remain relatively small, which means that new editions inevitably bring with them both deletions and additions. Indeed, any given addition may necessitate that an existing entry be removed. Space is at a premium. And changes to the dictionary's contents are supposed to "reflect language as it is used," according to one Oxford University Press spokesperson.[52] If some words go out of fashion, the press's logic suggests that they should be excised in favor of those that continue to be "used." Far from being a neutral repository of age-appropriate language, then, the *Oxford Junior Dictionary* is one of the sites wherein significant, if also subtle, decisions are made about which words—and which names—are worthy of recollection, and so also which words—and which names—can be forgotten or excluded.

When Oxford released its revised *Junior Dictionary* in 2007, hardly anyone noticed what words had been cut. Absence, by its very nature, can be hard to detect. However, one "Lisa Saunders, a mother of four in Northern Ireland, first noticed Oxford's disappeared words in 2008. While helping her son with homework," Stefan Fatsis reports in the *New Yorker*, "she realized that 'moss' and 'fern' had gone AWOL."[53] And Saunders found other changes, too: the words "saint" and "psalm," among other religious terms, had disappeared. All in all, in addition to the spiritual terms, the names of dozens of natural things had been deleted. "The deletions," Robert Macfarlane notes in *Landmarks*, "included *acorn, adder, ash, beech, bluebell, buttercup, catkin, conker, cowslip, cygnet, dandelion, fern, hazel, heather, heron, ivy, kingfisher, lark, mistletoe, nectar, newt, otter, pasture,* and *willow*." The common names for a myriad of plants, animals, and places, in short, dropped from the dictionary's pages. The additions were more conspicuous. As Macfarlane notes, "The words introduced to the new addition included *attachment, block-graph, blog, broadband, bullet-point, celebrity, chatroom, committee, cut-and-paste, MP3 player,* and *voice-mail*."[54] Where "ash" (tree) once stood, attachment had, like an invasive sapling, now taken over. Commentators like Macfarlane, although understanding of the editors' predicament, lamented the loss of so many names for natural things, particularly given that it was words for human-made technologies that replaced them. That the former was forced to cede ground to the latter is just one more example, critics complained, of the ways in which technologically advanced societies tend to destroy the natural world. As Macfarlane put it, "what is lost along with this literacy is something precious: a kind of word magic, the power that certain terms possess to enchant our relations with nature and place."[55]

This minor crisis, which one journalist described as a "lexicographic kerfuffle,"[56] attracted significant public attention in early 2015 when Margaret Atwood (author of *The Handmaid's Tale*), Helen Macdonald (author of *H is for Hawk*), Macfarlane, and twenty-five other well-known writers, scholars, and journalists, submitted a public letter to editors at Oxford University Press encouraging them to reinstate the names of the natural things in a future edition of their *Junior Dictionary*. The signatories claimed to be "profoundly alarmed" by the loss of so "many words associated with nature and the countryside." They worried especially that the words added to the dictionary implicitly approve of the "interior, solitary childhoods of today," lives spent largely in front of one screen or another, lives spent playing video games rather than playing in the woods, lives spent adjusting to the newest technologies instead of wondering at the endurance and the transformation of the natural world. And they criticized the *Junior Dictionary*'s editors for dodging their duty to "shape children's understanding of the world, not just to mirror its trends." The letter's signatories recognized the entanglement of word and world, of names and our capacity to know and to care for that which exceeds the human. Language, they claimed, shapes our understanding of our relation to the natural world and, therefore, how we treat our earthly cohabitants and the places where we dwell. If we lose our language for the more-than-human, they warned, we risk losing our connection to what exceeds the human, too.[57]

To recover the names of our earthly cohabitants, then, is to both shore up our sense of connection with what exceeds the human and to reckon with who and what might yet be lost. If we do not know the names of these others, we will have a hard time recognizing if they are disappearing. Recovering these names is a way of recovering our ability to grieve ash, beech, conker, and so forth. In the wake of the *Oxford Junior Dictionary*'s "lexicographic kerfuffle," Macfarlane and the visual artist Jackie Morris set to work on a large-format book joining together words and images called *The Lost Words: A Spell Book*. As Macfarlane and Morris explain in their single-page introduction:

> You hold in your hands a spellbook for conjuring back these lost words. To read it you will need to seek, find, and speak. It deals in things that are missing and things that are hidden, in absences and appearances. It is told in gold—the gold of the goldfinches that flit through its pages in charms—and it holds not poems but spells of many kinds that might just, by the old, strong magic of being spoken aloud, unfold dreams and songs, and summon lost words back into the mouth and the mind's eye.[58]

What follows are twenty entries, from *acorn* to *wren*. For each name, Macfarlane has crafted an acrostic spell song; each line begins with a letter that spells out the lost name. Here, for example, is the spell for ivy:

> *I am ivy, a real high-flyer.*
> *Via bark and stone I scale tree and spire.*
> *You call me ground-cover; I say sky-wire.*

Each spell is accompanied by Morris's evocative watercolors depicting the ways of being on earth that we lose touch with when we lose their names. To the right of each spell song lies an image of the named animal or plant, while, on the subsequent spread, Morris presents a *tableau vivant* which situates the being in a wider scene. Together, Macfarlane's words and Morris's paintings lead readers to the words and world around them. To read *The Lost Words* is to (re)learn the common names of twenty relatively common—and, hence, often overlooked—cohabitants of the British landscape, to speak those names out loud (for the poems demand to be read aloud), and to connect those names with "faces." It is, in other words, to become (re)acquainted with the teeming more-than-human earth that our screens all too often obscure.

In some significant sense, losing the words we use to name what exceeds the human means depriving ourselves of a rich—and enriching—bond with beings and ways of being with whom we share the earth, and with the earth itself, with the landscapes and places we inhabit. "As we further deplete our ability to name, describe, and figure particular aspects of our places,"

Macfarlane warns in *Landmarks*, "our competence for understanding and imagining relationships with non-human nature is correspondingly depleted."[59] Learning to name plants and animals, places and landscapes, offers a "way in" or a "way back" to those bonds that interiorized, screen-based lives quietly sever. Consider *Home Ground: A Guide to the American Landscape*, a compendium of hundreds of terms used to describe specific features of diverse landscapes. Assembled by the author Barry Lopez, *Home Ground* comprises entries spanning the comparatively familiar "ford," a "shallow place in a river where a man or an animal can cross by wading,"[60] to the less common "fosse," a "long narrow depression that lies between the front of a moraine and its outwash plain."[61] Perusing *Home Ground*'s pages, one encounters other enticing names like "antidune" and "brûlé," "comb ridge" and "dead ice," "ecotone" and "fumarole." Similarly, in *Landmarks*, Macfarlane gathers together some of the hundreds of names given to features of the British landscape over the centuries. In ten separate glossaries, he records names such as the Gaelic "bugha," a "green bow-shaped area of moor grass formed by the winding of a stream,"[62] and the native Somerset word "zebn-slaper," one name for the tiny, adorable dormouse (*Muscardinus avellanarius*).[63] Thus, *Home Ground* and *Landmarks* alike recover and reinvigorate hundreds of names, appellations which wake us up to the places where we dwell.

The more acutely attuned we are to the world around us, the better equipped we are to notice transformations, including losses, deaths, disappearances. By presenting readers with an incredibly precise, particularizing vocabulary of names, projects like *Landmarks* and *Home Ground* invite readers to see, more fully, what is present in the world around them and to feel, more intensely, the specificity of what they might lose. The names that populate the pages of these two books help us all to see that it is not just undifferentiated landscapes at stake, but a vibrant, diverse array of distinct beings, ways of being, and places—many, though not all, with names—that find themselves under threat of erasure. Learning these names can help us hone our perception. We need this sort of particularizing language if we are to recognize the real scope of what could be lost. Indeed, if we are to have any hope of staving off ecological and earthly losses, we must be able to name what is on the line and whose lives are at stake.

Increasingly, though, we are realizing that "what is on the line" is much more than specific plants or animals or habitats or ecosystems. It is the earth itself that is threatened today. Or, more precisely, it is the earth as humans have always known it that is at stake now. The earth systems that have supported life as we know it are coming undone as a result of some humans' actions, including centuries of burning ancient, finite fossil fuels. We have long known that humans shape their environments: we dig, we plant, we plow, we harvest, we build, we consume. We transform the world we live in. However, in the wake of decades of research, we now know that humans also "play a significant role in shaping the *Earth*, that is, the Earth that evolves as a totality, as a unified, complex system comprised of the tightly linked atmosphere, hydrosphere, cryosphere, biosphere, and lithosphere."[64] These are "earth systems," not local environments, and human actions are altering them, perhaps permanently, in ways that we are only starring to comprehend. For instance, humans once believed—because it was once true—that the weather is beyond human influence. Humans can do many things, we thought, but change the weather is not one. In an age characterized by anthropogenic climate change, however, we are compelled to acknowledge that, at last, even the weather has come under our sway.

The earth that we are shaping is not the earth we evolved to inhabit. So fundamental are the changes that the author and environmentalist Bill McKibben, best known for writing *The End of Nature*, the first popular book on climate change in the late 1980s, has insisted we need to rename the earth. His proposal? *Eaarth*.[65]

If McKibben is right that we need a new name for this planet we call our home, we need to think carefully about how that new name might reshape our collective relation with the most basic conditions of human life. The extra "a" in Eaarth—visible when written out but inaudible when spoken aloud, reminiscent of Derrida's *différance*—renders the planet strange.[66] That extra "a" subtly suggests that something is off kilter. McKibben's "Eaarth" is, as Paul Huebener puts it, "an intentionally uncanny sign of our irrevocably altered world."[67] It draws our attention to the strangeness of the present moment.

But "Eaarth" also looks like a typographical error, a slight misspelling. (As I write these words, Microsoft Word automatically underlines Eaarth with a squiggly red line.) And that may be part of the problem with Eaarth. It's too close to what we are used to, too similar to earth, too familiar. Not uncanny or weird enough. It intimates a difference, but only a seemingly slight, accidental one. It suggests—like Word's nagging red line—that the error can be (quickly) remedied. Delete one letter and we are back to our old friend, our only home, earth. To be sure, McKibben's argument is not that the transformations are inconsequential or that there are easy solutions. Quite the contrary. For McKibben there is no deleting the extra "a" precisely because there is no going back to the earth as it once was. We might eventually dwell on Eeaarrtthh, but we can never live on earth again. And yet, on its own, the appellation McKibben proposes is neither different enough nor strange enough to adequately underscore the scale and intensity of the changes that the earth is undergoing. Indeed, the addition of the extra "a" may signal that there is *more* (not *less*) than there once was. Not subtraction, but addition. Eaarth, not Erh. And so, we need a name that more forcefully captures the tremendous losses, as well as the profound diminishment, the earth continues to endure.

In 2000, the atmospheric chemist Paul Crutzen and the biologist Eugene Stoermer proposed not that we rename the earth but that we recognize a new period in earth history, a period they dubbed the "Anthropocene." In a brief report in the International Geosphere-Biosphere Programme's *Global Change Newsletter*, Crutzen and Stoermer explain that the conditions that characterized the Holocene (the most recent period of earth history lasting ten to twelve thousand years, marked by a stable climate, and in which all of human history has unfolded) have been sufficiently disturbed to warrant the conclusion that we are now living in a new geological epoch. This new geological epoch, they claim, is the product not of inhuman forces but of distinctly human influences. Among the decisive transformations, Crutzen and Stoermer mention the rise in both human and cattle populations, the movement from rural to urban settlements, centuries of expansive and seemingly limitless burning of fossil fuels, degrading soils as a result of industrial-scale agriculture, intensifying pollution, and various modes of resource extraction (e.g., overfishing). "Considering these and many other major and still growing impacts of human activities on earth and atmosphere, and at all, including global, scales," Crutzen and Stoermer argued, "it seems to us more than appropriate to emphasize the central role of mankind in geology and ecology by proposing to use the term 'anthropocene' for the current geological epoch."[68] With this act of naming, Crutzen and Stoermer predicted the conclusion of one geological epoch and the beginning of another.[69]

Both within and beyond the sciences, the "Anthropocene" appellation has attracted widespread criticism. I wish to mention two aspects of this criticism here. First, some critics rightly maintain that the Anthropocene concept unduly erases differences among humans and obscures the ways that some humans—and, more precisely, some ways of organizing human life—have contributed more to the disruption of earth systems than others. As Métis intellectual Zoe Todd reminds us, "Not all humans are equally implicated in the forces that created the disasters driving contemporary human-environmental crises."[70] In highlighting *anthropos*, the Greek word for "man" or "human," as the guilty party in creating this crisis, critics contend, the Anthropocene concept covers over differences that make a difference.[71] We need to be able to distinguish between multinational fossil fuel corporations and Native American communities, for instance.[72]

Second, critics contend that the Anthropocene appellation does more to reinforce than to challenge human hubris and domination. Eileen Crist argues, for example, that,

> The [Anthropocene] discourse subjects us to the time-honored narrative of human ascent into a distinguished species; a naturalized, subtly glamorized rendition of the "I" as on par with stupendous forces of Nature; a homogenized protagonist named "the human enterprise" undefended for either its singularity . . . or its insularity; . . . a reification of demographic and economic trends as inescapable, leaving the historically constructed identity of *Homo sapiens* as planetary ruler undisturbed and giving permission to humanity's expansionist proclivities to continue—under the auspices of just-the-facts—as the independent variables of the situation; a sidestepping of confronting Life's unraveling, representing it instead as a worthy criterion of a new name; and a predilection for managerial and technological solutions, including a partiality for geoengineering, which, if worsening climate scenarios continue to materialize, will likely be promoted as necessary to save civilization.[73]

As Crist makes clear, the Anthropocene appellation risks bolstering the very arrogance and anthropocentrism of some humans and, thus, of further damaging relations among humans and the earth.

"Anthropocene" is not the only name in town, however. In the wake of Crutzen and Stoermer's publicizing of the name in the early 2000s, scholars have proposed other names for the time we are living in. In fact, according to one study, more than ninety other names have been recommended, among them "Androcene" and "Anglocene," "Entropocene" and "Eurocene," "Homocene" and "Homogenocene," "Metropocene" and "Misanthropocene," "Plantationocene" and "Plastocene," "Technocene" and "Trumpocene."[74] Alternate names identify alternate sources for the present planetary transformations, from men to Europeans to plantation culture to Donald Trump's particular brand of climate denialism; identifying different culprits, of course, suggests different ways forward. Coined by the environmental historian Jason Moore, "Capitalocene" is perhaps the most popular of the alternative names for our present epoch: as its prefix suggests, "Capitalocene" views capitalism—understood as "a system of power, profit, and re/

production in the web of life"—as the basic cause of our global ecological crisis.[75] Despite this proliferation of alternatives, it is "Anthropocene" that continues to stimulate scientific and popular interest. Hence my focus on it here.

Like the proper names of endlings, the common and Latin names of other species, and the names we bestow upon places and various features of the landscape, the name "Anthropocene" also foreshadows a kind of mourning. We hear in the name Anthropocene an echo of two ancient Greek words: *anthropos* and *kainos*. On the one hand, *anthropos* means "man" or "human." Hence *anthropo*logy, the study of human beings, and *anthropo*centrism, the belief that human beings are different from and better than all other kinds of beings because we possess reason, the capacity for meaningful speech, or some other reason. On the other hand, *kainos*—from which the suffix *-cene* is derived—means "recent" and "unprecedentedly new." *Kainos* is one of two Greek words for "new." The other is *neos*, from which we receive words like *neo*conservative and *neo*liberal. While *neos* means "new" in the sense that one can have a new version or iteration of something which already exists (neoliberalism is, after all, an extension of classical liberalism, not a radically new political orientation), *kainos* suggests originality, novelty. And so, when we call the current geological epoch Anthropocene, we are describing an unprecedented time, a time unlike any other that has come before. We are also conjuring a notion of humanity—of who *we* are—that is without precedent. Anthropocene means both the "new age of the human" and the "age of the new human."

As even this brief etymological detour suggests, the name Anthropocene signifies something radically, unprecedently new. It suggests, too, that we have left, or are leaving, something behind. That we have lost, or are losing, something. And, thus, that the work of mourning awaits us. But what sorts of loss does the name Anthropocene intimate? The two definitions of Anthropocene mentioned earlier—"new age of the human" and "age of the new human"—indicate at least some possibilities.

The first description ("the new age of the human") implies that we are living amid a new period of earth history characterized by human influence, which also means that we have left the previous geological epoch behind. What do we lose in doing so? Abandoning the Holocene, we abandon the conditions under which human life, and human civilization, have evolved. We take leave of the only earthly conditions humans have ever known. The Holocene was marked by a relatively stable and predictable climate, and it was this climate that fostered the growth and development of *Homo sapiens*. The name Holocene is derived, like Anthropocene, from two Greek words: *holos*, meaning "whole," and the now familiar *kainos*. Holocene means, plainly, "wholly new." From an anthropocentric perspective, part of what made the Holocene entirely new was the relatively rapid growth and expansion of human beings and the corresponding advent of myriad intellectual, technological, cultural, and political developments. Since all of our know-how was cultivated during the Holocene, moving into the Anthropocene also means that we are losing our bearings and moorings. Indeed, some believe that we are losing civilization as we know it.[76] Our old truths may well no longer hold in this new world.

The second definition implies the breakdown of one such "truth," namely our taken-for-granted notions of what it means to be a human being: not only have we disrupted the earth systems that have always supported human life but, in the process, we have become something other than what we used to be. By collectively shaping the earth, we have also reshaped what it means to be human: to the list of our characteristics and responsibilities we must now add our capacity to make and unmake the earthly conditions that undergird our very being, not to mention the conditions that shape the lives of our more-than-human cohabitants. In a sense, then, we are also losing our sense of who we are as *Homo sapiens*. If ever we believed that humans are incapable of interfering with or altering the earth itself, we must abandon that belief immediately. This means that we must also leave behind that vague sense of innocence that has allowed so many of us to ignore our own roles, however minor, in disrupting earth systems. In other words, the name Anthropocene prepares us to mourn, as Nancy Menning has put it, "not only what we have lost, but also what we have destroyed."[77] The name Anthropocene invites us to see ourselves, perhaps for the first time, as destroyers, as unmakers of worlds that have taken eons to form. It inaugurates us into a kind of species thinking, which asks us to see ourselves not as discrete, autonomous, free-willed individuals but rather as representatives of the species *Homo sapiens*, which exerts its own force over and beyond any intended, conscious, willful decisions we might make. It is an uncomfortable realization but a necessary one.

So, it seems that the name Anthropocene calls us to grieve in a doubled sense: to grieve the stability and predictability of the Holocene, the end of which *Homo sapiens* has brought about, and to grieve the relationship we once had with the earth, a relationship in which humans were governed by conditions not of our own making. There is a third sense, though, that should be mentioned. The name Anthropocene also gestures toward a time to come—a time after the Anthropocene, if not also after the *anthropos*. As Derrida has taught us, every name foreshadows a loss to come. Even as it declares the emergence of a novel period in earth's history, "Anthropocene" simultaneously suggests the impermanence, in geological terms, of our own time. This name anticipates, already and at the very beginning of this "new age," the end of this geological epoch. It says that a time will come when this relationship between *Homo sapiens* and the earth will wither away, when human beings will either cease to cohabit the planet or finally make ourselves at home as part and parcel of the land and earth communities to which we belong. The name asks us, therefore, what we are willing to give up—what we are willing to lose—so that we do not lose it all. Indeed, the name Anthropocene asks us to reckon with, to prepare ourselves for, the losses to come: the loss of civilization as we know it, the loss of the ideal of everlasting progress, the loss of human centrality and superiority, the loss of economic self-interest as the only possible metric on which to base decisions, the loss of a vision of the future that looks like the recent past. If we wish to dwell more peacefully on earth with others, human and more-than-human alike, such a reckoning seems both unavoidable and inherently worthy of our attention.[78]

Walking the Bellefonte Rail Trail, day after day, I am slowly but surely getting to know the local flora and fauna. As my partner and I saunter and share news about our days with one another, I frequently stop to investigate the flowering plants that catch my eye or to identify the birds and other animals whose paths we cross. Sometimes under my breath, sometimes aloud, I call them by their names: evening primrose, American goldfinch, prairie fleabane, gray catbird, jewelweed, groundhog, swamp milkweed, northern cardinal, Joe-Pye weed, white-tailed deer, goldenrod, monarch caterpillar, cutleaf teasel. When I find someone or something whose name I do not yet know, I take a photograph or two. Back home, trying to put a name with a form, I flip through my guidebooks or search the internet. Each name I learn is a link forged with the plants and animals with whom I share the earth, a linguistic bond that makes it that much easier to appreciate who else inhabits this land community. Reciting the names, I feel a little bit closer, a little more intimately connected, to the more-than-human others in my midst. As these names roll off the tip of my tongue, I cannot help but feel that learning to utter these names is a small yet certain way to usher in something other, something better, than the Anthropocene.

CHAPTER 3

Revealing Loss

On Archiving

Knowledge and Sorrow

On the final day of June, while the sun was setting, Brandon and I finished supper and set out for our evening amble along the Bellefonte Central Rail Trail, a footpath following the abandoned rail line that once connected State College to the nearby borough of Bellefonte. When we left our house, the air temperature was 75 degrees Fahrenheit, and a light breeze kept us cool as we strolled beneath the canopies of the oaks, maples, elms, and evergreens lining the edges of the path where native grasses and flowers thrive alongside "invasive" species like the ubiquitous broadleaf plantain (*Plantago major*), known to some as "White Man's Foot" because it appeared to pop up everywhere European settlers travelled. Groves of staghorn sumac (*Rhus typhina*) were blooming, too, their fuzzy auburn drupes suspended like red lanterns on the ends of long, lanky limbs spangled with slim football-shaped leaves. Overhead, the occasional cumulous cloud punctuated the otherwise cerulean sky that had persisted for days since the last round of thunderstorms dumped rain on the valley. Almost four inches of rain fell in June, a bit more than usual, but a bit less than last year. When we set out for our walk, everything was growing and green.

Yellow, however, was to be the theme of the night. We had not walked fifty paces before a flock of American goldfinches (*Spinus tristis*) playfully whirred overhead. One of the bright yellow birds narrowly missed hitting the side of my head, so low did the *chip-chip-chipping* crew descend at one point before zipping into the sumac at breakneck speed. After watching the goldfinches collectively swoop in and out of the canopy above us and into the meadow on the other side of the path for a while, we walked on. Before long we halted again, this time to examine the flowers which had, in a few days' time, transformed the meadow from an overwhelmingly green canvas into a multihued painting. Sprawling swathes of crown vetch (*Securigera varia*) covered much of the meadow during the summer months. When we went for our walk on June 30, though, the *Securigera*'s pink and white blossoms, which had already been in bloom along roadsides, railways, and other forgotten places for weeks, were broken up by the brilliantly yellow flowers of hundreds of lance-leaved coreopsis (*Coreopsis lanceolata*) newly in bloom. From afar we supposed they were black-eyed Susans (*Rudbeckia hirta*). Up close, their yellow centers betrayed their identity. Closer to the ground, patches of birdsfoot trefoil (*Lotus corniculatus*) sent their distinct golden blooms sunward. And hundreds of small flowers with thin white petals and yellow centers also mottled the meadow. Prairie fleabane (*Erigeron strigosus*) resembles its better-known, larger cousin, the common daisy (*Bellis perennis*). We also saw reminders of yet another yellow flower—*Taraxacum officinale*—while walking along. Once the common dandelion's flower head has died, dried out, and dropped to the ground, its distinctive puffy, white seed head unfurls to scatter seeds every time the wind blows and every time someone plucks one out of the earth and blows it out like a birthday candle. These spectral blowballs promise still more golden flowers in the future.

So go most of our evening walks these days. We spend as much time hunching over flowers and squinting at birds as we do walking; we expend as much effort attending to the more-than-human beings with whom we share this sliver of earth as we do ambling. Indeed, our walks are as much about exercising as they are about grounding ourselves in this place, about getting to know the earth beneath our feet. It is a matter of attention, of learning to attune to the myriad of beings and ways of being around us, of noticing the subtle rhythms and patterns unfolding in our midst. It is a matter of unlearning some of the habits of inattention cultivated in an increasingly virtual, air-conditioned, interior world—a world in which the differences among black-eyed Susans and lance-leaved coreopsis are irrelevant; a world in which the weather, much less the season, hardly seems to affect our everyday lives; a world in which most of us pay no attention to what is blooming, to what is buzzing, to who is coming, to who is going; a world in which we remain largely oblivious to what Aldo Leopold once called "the hundred little dramas of the woods and meadows."[1] It is not a question of choosing between two worlds: one real, one virtual, or one natural, one artificial. After all, when we finally make it home from our walks, it is to field guides and websites and computer applications that I turn to help me identify and understand who and what we have seen. Sometimes I share my observations through photographs and text on Facebook and Instagram. In doing so, I generate a small record—a written trace in an archive always in the making—of the more-than-human goings-on at a particular time and in a particular place. This is a written record, too, of who and what are at stake, of who and what may one day be lost.

There is a name for these entwined practices of observation and documentation, witnessing and archiving: *phenology*. Coined in 1849 by the botanist Charles François Antoine Morren at a meeting of the Belgian Royal Academy of Sciences, phenology was initially described as "a specific science which has the goal to know the *manifestation of life ruled by time*."[2] Sometimes called "nature's calendar," phenology is concerned with cyclical and seasonal natural phenomena—earthly events that occur with regularity and whose timing is affected by climate. Phenological events include, among numerous others, the arrival and departure of migratory birds, the break-up of ice on frozen rivers and lakes, the first time a flowering plant blooms, the beginning and end of hibernation periods, the onset of senescence in trees and other plants, and the first and last frosts of the year. Each and every biotic and abiotic process has its own phenophases, which can be detected and documented.

Phenology, whether practiced by professionals or laypersons, involves witnessing, recording, and archiving the goings-on of plants and animals, of habitats and ecosystems, of weather and climate. It demands not only that we perceive seasonal events but also that we generate a written account of their occurrence, too. It is thus that phenology revives some of the earliest functions of writing: inventorying what is and accounting for what happens.[3] Phenology is a method of accounting for what happens in the more-than-human world from one day, one season, and one year to the next. "The events comprising the annual cycle are innumerable," Aldo Leopold and Sara Elizabeth Jones write in the opening pages of their report on a decade-long phenological investigation of central Wisconsin's flora and fauna. "From the beginnings of history," Leopold and Jones continue, "people have searched for order and meaning in these events, but only a few have discovered that keeping records enhances the pleasure of the search, and also the chance of finding order and meaning. These few are called phenologists."[4]

With varying levels of systematicity, people have indeed sought to grasp underlying ecological principles by attending to, recording, and studying seasonal events. The practice is both ancient and enduring: "Phenology, or something like it," Mark Hineline writes, "is doubtless old as desert dust."[5] Hunter-gatherers, dependent upon plants and animals whose presence shifts with the seasons, surely attended to what we would now call phenological events. So, too, with farmers: it helps to know when the first and last frost of the year occur, for example. Sow your seed too early and all your hard work may be for naught. Attending to and remembering these events was, for much of our history, a matter of survival.

Indigenous peoples have long incorporated aspects of what we now call phenology into sophisticated ways of knowing that comprise cyclical conceptions of time and in which Indigenous, cultural, and spiritual knowledges sit alongside and enrich forms of ecological and seasonal knowledge rooted in lived experience.[6] As one group of researchers explains, the timing of seasonal phenomena is an important aspect of how many Indigenous people conceive of their relations with and responsibilities to other beings and ways of being. For instance, an elder of the Confederate Tribes of Siletz Indians told these researchers about how his community historically relied on the presence of carpenter ants to instruct them on when to start hunting eel; although the carpenter ants and eels are not directly connected to one another, they had long appeared at the same time of year such that the presence of one indicated the presence of the other.[7] To mention just one other example, for Indigenous peoples who have lived near Lake Oneida in present-day New York, the blooming of a particular flower has served as a cue "that the ground has thawed, ploughing can begin, burial services can be performed for those who have died during the winter, a particular fish is running in the river, and traps can be set for small animals that have emerged from hibernation."[8] As these two examples show, natural events set into motion cultural events and practices. Or, in other words, the timing of natural events instigates particular, situated relations among human and more-than-human others. In the repetition and renewal of these relations from season to season lies the coherence of a culture. Whether transmitted orally or through writing, the remembrance of natural events and the relations they initiate form what Karim-Aly S. Kassam and colleagues call "ecological calendars," which they describe as "knowledge systems to measure and give meaning to time based on close observation of one's habitat."[9]

If an Indigenous community's ecological calendar—their collective sense of when natural events happen and, therefore, when they should begin or end particular practices—is disturbed in some way, the community may find itself lost or adrift. "Disruption of the seasonal cycle . . . has profound effects for Native cultures," explains Samantha Chisholm Hatfield and colleagues in their work on "Indian Time." This is so, they contend, because disruptions to ecological calendars "challenge the fundamental belief about how elements of the natural world are connected, as well as the timing of when traditional patterns occur and behaviors are performed."[10] When an event fails to occur, or fails to occur when people expect it to, something is surely lost. Sometimes the natural event itself is lost; such is the case when a species that once migrated to an area no longer arrives. Sometimes a people's sense of place—of what it means to inhabit a place—erodes. When, for instance, seasonal events happen sooner or later than in the remembered past, one might feel that one has lost one's bearings among more-than-human others and in a place.

Several prominent Americans closely associated with environmentalism have also practiced phenology, albeit it for somewhat different purposes. In the mid-1700s, for instance, a young Thomas Jefferson kept a "Garden Book," in which he noted the flowering dates of plants on his family's farm in Virginia.[11] Later, as the third president of the United States, Jefferson instructed Meriwether Lewis and William Clark to document "the dates at which particular plants put forth or lose their flowers, or leaf, times of appearance of particular birds, reptiles, or insects" on their westward expedition.[12] (Of course, while they gathered information about the flora and fauna of the United States' newly acquired lands west of the Mississippi River, Lewis and Clark also laid the groundwork for the further dispossession of Native peoples.) A century later, in the years leading up to the American Civil War, Henry David Thoreau kept meticulous written records of hundreds of natural cycles near his beloved Walden Pond.[13] Thoreau's goal was to construct an exhaustive "Kalendar" of phenological events in Concord, Massachusetts.[14] With the same verve that Thoreau approached his own phenological observations, Leopold and Jones documented hundreds of seasonal events in central Wisconsin between 1935 and 1945.[15] And Rachel Carson, though less systematic than either Thoreau or Leopold in documenting natural events in daily life, incorporated phenological knowledge into much of her writing. Indeed, the title of her best-known book, *Silent Spring*, hints at the breakdown of familiar seasonal events.[16]

Precipitated by the intensification of climate change, today thousands of people are practicing phenology in their backyards, neighborhoods, and ancestral lands. In places all over the world, people are noticing and noting, perceiving and inscribing, witnessing and archiving—in notebooks and online databases and Instagram and beyond.

Phenology, like naming, prepares us for the work of ecological mourning. As I will demonstrate in the following pages, phenology discloses ecological losses which we might otherwise miss. Phenology's entwined practices of witnessing, recording, and archiving reveal three different yet connected kinds of losses—those that are *happening now*, those that have happened in the *past*, and those that will have happened in the *future*. Observing seasonal phenomena foments familiarity with who and what are present in an ecosystem at a given time of the year. Through careful observation, phenologists become attuned to what is going on in the land community to which they belong. And the more attuned we become to what is present in a specific place, the likelier we are to notice when plants and animals go missing, when habitats or ecosystems take a turn for the worse, when enduring rhythms begin to break down. Recording these phenomena—which is to say, creating a written trace of such events—enables us to look back and to compare the present with the past. These records shed light on continuity as well as change, on persistence as well as loss. And archiving these year-to-year records lays the foundation for an ecological grief *to come*, a grief whose form we can anticipate, but whose object reveals itself only in time. When phenological records accumulate over extensive periods of time, they can reveal long-term transformations that are difficult, often impossible, to discern at any given moment: earlier spring flowering and migrations, for instance, and the local extinction of flora and fauna. Although formed in the past, phenological archives direct our attention to the future, to those species threatened with erasure. Phenological archives also become the condition of possibility for telling new stories about ongoing ecological disturbances. As we live into a geological epoch marked by climate change, phenological records of the past furnish us with the inventional resources that we need for telling the story of a present and a future characterized by still more loss.

In the next part of this chapter, I contemplate the ways in which phenology, if we understand it as a fundamentally archival and rhetorical practice, prepares us to grieve for the more-than-human beings we are already losing and for those threatened with erasure in the future. Before engaging phenology at the conceptual level, however, I wish to dwell on a representative anecdote that can indicate both the thrust and the stakes of this argument.

"One of the penalties of an ecological education," Aldo Leopold once wrote, "is that one lives alone in a world of wounds."[17] The more we know about the earth and about the more-than-human others with whom we share the earth, Leopold reasoned, the more apt we are to recognize signs of distress and symptoms of imbalance. The more perceptive we become of the land's health, that is, the more capable we become of noticing, and of being affected by, losses in the land community. In Leopold's remark we can hear an echo of a still more ancient intelligence embedded in Ecclesiastes 1:18, which reads: "For in much wisdom is much grief; and he that increaseth knowledge increaseth sorrow."[18] Likewise, in Leopold's phenological practice—and its enduring legacies—we can glimpse some of the ways that wisdom and grief, knowledge and sorrow, intensify one another.

Leopold's interest in phenology deepened after he and his wife, Estella, bought a worn-out farm about an hour's drive northwest of Madison, where he taught game management on the University of Wisconsin's flagship campus.[19] In the foreword to his now-classic *A Sand County Almanac*, Leopold describes the eighty-acre farm along the Wisconsin River as his "week-end refuge from too much modernity," a place where he would "try to rebuild, with shovel and axe, what we are losing elsewhere."[20] By the time the Leopolds acquired the property in 1935, at a rate of eight dollars an acre, the land had been heavily worked and severely diminished.[21] Approaching the property for the first time in January of that year, Leopold surveyed a snowy, rundown scene:

> Clumps of aspens sprouted along the margin of a frozen marsh. A haggard row of wind-swept elms lined a driveway. The fields, poor and sandy even in

> the summer, seemed even more barren when gripped by winter. The spent soils supported only sand burrs, their dried heads held out above the snow. The farmhouse at the end of the elms had burned down; all that remained of the house was a dug-out foundation. To one side there was a small but sturdy chicken coop, the only structure [which was] still standing on the property.[22]

The Leopolds converted the coop into sleeping and cooking quarters and built an outhouse, jokingly named the Parthenon, which still stands. Although imagined as a hunting camp, the farm became a living laboratory, a place where Leopold could labor and learn; a place where he could exercise his observational muscles upon the visual, auditory, and olfactory signals emitted by the flora and fauna of central Wisconsin. (The land is maintained by the Aldo Leopold Foundation, which preserves the 600 acres the Leopolds eventually amassed as an educational center.[23]) And it was there on the farm that Leopold would apply himself most fully to the work of phenology.

Even before he and his family rolled down the dirt road that leads from the highway to the farm, though, Leopold noticed seasonal changes in the surrounding landscape. In the "July" chapter of his *Sand County Almanac*, in an underappreciated essay entitled "Prairie Birthday," Leopold recounts one especially illustrative experience. "Each July," he writes, "I watch eagerly a certain country graveyard that I pass in driving to and from my farm."[24] Among the headstones and the lovingly placed flower bouquets, Leopold discovered "a pin-point remnant of the native prairie on which the graveyard was established" more than one hundred years earlier. Having long been spared, first from the scythe's blade and then from the mower's, a forgotten corner of the triangular cemetery sheltered wildflowers and grasses by then uncommon in more "developed" areas. It was in this "yard-square relic of original Wisconsin" that he spotted a singular "stalk of compass plant or cutleaf Silphium, spangled with saucer-sized yellow blooms resembling sunflowers."[25] *Silphium laciniatum* grow straight up, their stalks stretching toward the sun, occasionally reaching heights of ten or more feet. Bumblebees love them. And so did Leopold. For at least seven years, he returned each

July to examine this singular specimen. "This year," he reports, "I found the Silphium in first bloom on 24 July, a week later than usual; during the last six years the average date was 15 July."[26] Here, Leopold condenses into just twenty-eight words seven years' worth of data on this plant—the culmination of seven years of visiting the *Silphium* each summer, seven years of noticing when it first bloomed, seven years of taking the time to document those dates, and seven years of keeping his records in a safe place where they could later be retrieved. Condensed into those twenty-eight words, then, are the fruits of a seven-year-long relationship between one person and one plant.

Year after year, Leopold returned to witness the blooming of this single specimen of *Silphium laciniatum*. This fact alone goes a long way towards explaining his heartbreak when, just a few days later, he discovered that the plant had been destroyed. "When I passed the graveyard on 3 August," Leopold laments, "the fence had been removed by a road crew, and the Silphium cut." The graveyard was, thus, forced to accept still another body—this one floral but not, for that reason, any less significant for Leopold. Indeed, he saw it as an indication of trouble on the horizon. This event, he worried, anticipated a cascade of further losses: "It is easy now to predict the future; for a few years my Silphium will try in vain to rise above the mowing machine, and then it will die. With it will die the prairie epoch." In the destruction of this *Silphium*, in other words, Leopold detected the disappearance of the prairie, of an ecosystem encompassing a myriad of other species similarly relegated to that most unsavory category: "weed." The problem, he suggests, is that too many of us know too little about the plants with whom we share the earth. And in our ignorance, we feel no pain when our floristic cohabitants wither—or are pushed—away. "Mechanized man, oblivious of floras," he writes, "is proud of his progress in cleaning up the landscape on which, willy-nilly, he must live out his days."[27]

While this generic "mechanized man" cannot see the native flora's value, Leopold's sustained engagement with the *Silphium* leads him to cherish the plant and so also to grieve its loss. Drawing upon his own experience, he conjectures that, "We grieve only for what we know."[28] Thus do knowledge and sorrow, wisdom and grief, arrive in tandem. The more we know about our earthly cohabitants, the more intimately engaged we are with their lives and ways of life, the more intensely will we feel their loss. Bereavement literally means to be "deprived of a close relation through death." A sense of deprivation emerges only when we feel "close" to the lost other. For Leopold, that feeling of "closeness" with the *Silphium* was surely a product of his phenological practice. Had he not returned, again and again, to check in on this plant he never would have noticed its untimely demise. Observing what is present exposes us to absences and disappearances. Correspondingly, he describes its loss in terms that express his sense of connection and loss, of entanglement and bereavement: "This is one little episode in the funeral of the native flora," he remarks, "which in turn is one episode in the funeral of the floras of the world."[29] Leopold's metonymic reasoning is inherent in most ecological thinking, which views all things as always already associated and interconnected.[30] And it is that same associational logic that drives Leopold to link the loss of a single plant in central Wisconsin to the fate of the world's floras. Loss, he suggests, begets loss. That may well be the lesson of the Anthropocene, too: thanks to feedback loops we do not fully understand, ecological and earthly losses cascade one atop another.[31]

Interestingly, Leopold connects the potential loss of *Silphium* to the near loss of the American buffalo (*Bison bison*) in the nineteenth century. This association is forged in two different ways in "Prairie Birthday." First, a pencil drawing of a buffalo appears at the start of this essay; drawn by Charles H. Schwartz, the illustration depicts a buffalo from the neck up in profile. The buffalo is nibbling on the stalk of a *Silphium* plant. Second, at the end of the essay, Leopold wonders at the persistence of *Silphium*, which he surmises can tolerate "almost any kind of mutilation except continued grazing, mowing, or plowing."[32] Whereas cattle in enclosed pastures often eat the compass plants until none remain, Leopold guesses, unfenced buffalo must have left some *Silphium* plants uneaten as they grazed from here to there.[33] Having made this connection, Leopold ends the essay with the following line: "Few grieved when the last buffalo left Wisconsin, and few will grieve when the last Silphium follows him to the lush prairies of the never-never land."[34] Bison, we must remember, did not "leave." They were, rather, systematically exterminated in an attempt to deprive Native communities of a key source of sustenance and cultural coherence.[35] Readers can suppose, then, that those "few" who grieved the bison's disappearance in Wisconsin were those who had established and maintained close ties to the animals, especially the Indigenous peoples for whom buffalo constituted a key part of their spiritual life and subsistence economy. But Leopold leaves this connection to the reader to form. Just as "mechanized man" could not notice the loss of *Silphium*, among other plants, Leopold suggests that early American settlers could not feel the loss of the buffalo *as a loss*, and not only because they enacted a violent campaign to drive the buffalo to and over the edge of extinction. Here again, close connection—whether via daily interaction or concerted observation—becomes a pathway to recognizing loss. The early settlers, dead set on conquering the frontier, lacked precisely what the Plains Indians possessed: a way of life in which connections with the buffalo were literally life- and culture-sustaining.

Thankfully, Leopold was wrong about the fate of *Silphium laciniatum*. It still grows today in more than half of the states in the Union, from New York to New Mexico. Though listed as "threatened" in Michigan and Tennessee and as "endangered" in Ohio,[36] *Silphium* mostly seems to have escaped the fate Leopold feared most: extirpation and then extinction. However, that is not to say that the species has escaped humanity's impacts upon the earth unscathed. As Sarah D. Wright and Nina Leopold Bradley (Aldo and Estella's third of five children) report in "Thinking Like a Flower," "During the 1930s and 1940s, compass plant began to bloom, on average, on July 15; today, it would no longer find its place within the 'July' chapter [of Leopold's *A Sand County Almanac*], as its average date of first bloom [between 1994 and 2004 was] June 26."[37] Drawing on phenological research conducted at the Leopold's farm between 1994 and 2004, Wright and Bradley demonstrate that *Silphium laciniatum* now blooms three weeks earlier than it did when Aldo Leopold was observing its annual growth and development in that roadside graveyard. As temperatures increase as a result of climate change, spring is literally arriving earlier in many locales.[38] "The study of phenology," Wright and Bradley explain, "exposes the unfolding saga of environmental change that we have unwittingly staged."[39] This "unfolding saga of environmental change" is characterized, as we saw in the Prologue, by incalculable ecological losses and earthly transformations. Revealing these losses and transformations is one of the unexpected side-effects of phenological practice.

I say "unexpected" because neither Leopold nor his phenological forebears foresaw the incredible transformations that anthropogenic climate change would eventually set into motion. Simply put, as Mark Hineline notes, "Leopold did not, himself, see climate change coming."[40] In the late 1940s, while putting the finishing touches on what would become *A Sand County Almanac*, Leopold understood that humans had already severely diminished many specific habitats and ecosystems, but the idea that humans could fundamentally alter something as inhuman as the climate was foreign. In fact, for Leopold and Thoreau alike, phenology "was a science running counter to apocalypse: instead of forecasting calamity or momentous shifts, the landscape's rhythms signaled environmental perpetuity."[41] Year to year, the dates of particular phenological events vary, but Leopold saw these variations as "data to be incorporated into an overarching pattern."[42]

Evidence of this perspective can be found on the first page of Leopold and Jones's 1947 article in the journal *Ecological Monographs*, "A Phenological Record of Sauk and Dane Counties, Wisconsin, 1935–1945," in which they report the dates of 328 distinctive phenological events over a decade: the skunk's annual emergence from hibernation, the northern cardinal's first song of the year, the arrival and departure of migratory birds, the accumulation and break-up of ice on the Wisconsin River, the first sighting of earthworms on the ground, the opening of silver maple buds, and the first and last frosts of the year, to mention just a few of the phenomena they recorded. The phenological tables alone run to nearly sixteen typeset pages. "Many of the events of the annual cycle recur year after year in a regular order," Leopold and Jones explain.

> A year-to-year record of this order is a record of the rates at which solar energy flows to and through living things. They are the arteries of the land. *By tracing their responses to the sun, phenology may eventually shed some light on that ultimate enigma, the land's inner workings.*[43]

In Leopold's experience, phenology confirmed the ongoingness of ancient earthly rhythms by showing that, from year to year, the same natural events take place at the same time. In his records Leopold mostly discovered cycles characterized by continuity and persistence: repetition without much difference.

Assuming a reasonably stable climate, over relatively short periods of time seasonal phenomena *do* happen at about the same time each year. However, in a period defined by climate chaos, continuity gives way to transformation, and persistence yields to disruption. Leopold's records provide us with a picture not of a pristine ecosystem in perfect working order but, rather, of a land community on the precipice of tremendous changes. In the seven decades that have elapsed since his investigation, decades some now describe as the "Great Acceleration," the human toll on the earth has increased exponentially. Since about 1950, the human population has risen from approximately two billion to nearly eight billion, and the population of cattle, pigs, and chickens (the animals Americans eat most often) has kept pace. In the same period, demands for energy (particularly coal, oil, and gas), water, and fertilizer have increased dramatically. While levels of carbon dioxide, nitrous oxide, and methane, among other greenhouse gases, were already climbing by the early 1950s, they have skyrocketed since Oxford University released the first edition of Leopold's *A Sand County Almanac* in 1949. As the International Geosphere-Biosphere Programme notes, "The second half of the 20th Century is unique in the history of human existence. Many human activities reached take-off points sometime in the 20th Century and sharply accelerated towards the end of the century."[44] Of course, the most conspicuous, and the most catastrophic, effect of the Great Acceleration is global climate change. And, as it turns out, climate change is reflected in the specific sorts of phenological events that Leopold studied in the 1930s and 1940s. Although Leopold believed his phenological observations might eventually "shed light" on the land's unchanging patterns, his records have, in fact, illuminated ecological "wounds" that he could not yet imagine at the end of his own life.

Completed immediately prior to the profound planetary changes of the past seven decades, Leopold's records now reveal a baseline against which more recent phenological observations can be compared. Between 1976 and 1998, for instance, Nina Leopold Bradley carried out phenological research at the Leopolds' farm, observing fifty-five of the same seasonal events her father had documented. Comparing her observations with her father's records revealed that "roughly one-third of the phenophases appeared to advance in earliness over the period, one-third appeared not to advance, and the remaining third were statistically intermediate."[45] Geese, red-winged blackbirds, robins, and cowbirds, to name a few, arrive earlier now than they once did. Hepatica, forest phlox, and black-eyed Susans bloom earlier now, too. As noted earlier, spring is literally arriving earlier today in central Wisconsin than it did in the mid-1900s. More recently, Leopold's phenological records have enabled scientists to discern that, from 1935 to 1945, the mean spring temperature in central Wisconsin was 7.5 degrees Celsius (45.5 Fahrenheit). Under those conditions, the mean flowering date for twenty-three phylogenetically distinct plants was May 7. By contrast, from 2002 to 2012, the mean temperature in the same place was 9.3 degrees Celsius (48.74 Fahrenheit) and the mean flowering date had jumped forward by an entire week.[46] The seasons as we once knew them are now withering away as climate change bears down upon places and land communities. As temperatures continue to rise, the seasons will continue to shift and transform. Such is the sad story that Leopold's phenological observations, preserved now in an archive at the University of Wisconsin and online, enable us to tell, to hear, and to feel.

I have dwelled on Leopold's case because it illuminates this chapter's central claim: the practice of phenology prepares us for the work of ecological grief in at least two senses. The practices of witnessing and recording draw us into intimate relations with our more-than-human cohabitants, relations which form the conditions of possibility for later feelings of loss. Had Leopold not visited the *Silphium* plant again and again to record its date of first bloom, it is unlikely that he would have noticed or been affected by its destruction. The intimate relationships forged through phenological practice in turn lead us to feel more entangled with our cohabitants. And this sense of entanglement is responsible for feelings of bereavement. Through phenology, we are able to overcome our "inability to really *get*... the multiple connections and dependencies between ourselves and these disappearing others" and to "appreciate all the ways in which we are at stake in one another, all the ways in which we share a world."[47] These practices of witnessing and recording, because they generate an archive to which we can return, also lay the foundation for recognizing losses that are currently unfolding and that might yet come to pass. By generating an archive of what appears, phenology shines a light on what disappears, too. Indeed, phenology reveals what we are losing now, what we have lost, and what we might yet lose.

Witnessing, Recording, Archiving

As noted earlier, the Belgian botanist Charles Morren coined the word *phenology* in 1849 to name that area of scientific inquiry concerned with the timing of seasonal natural events. The *Oxford English Dictionary* records a single definition of the word: "the field of science concerned with cyclic and seasonal natural phenomena, esp. in relation to climate and plant and animal life." This definition is surely correct, but it fails to capture a deeper and more significant dimension of phenology. Morren's neologism gathers two ancient Greek words: the prefix *phaino* and the suffix *logos*. Of course, *logos* means—simultaneously—word, reason, logic, study, and discourse. Appended to other words, the suffix *-logy* means "study of." From the less familiar *phaino*, we receive words like *phenomena*, meaning "that which appears," and *phenomenology*, meaning "the study of what appears (especially to human consciousness).” In one sense, *phaino* means “to appear” or “to become evident.” With this sense of *phaino* in mind, the word *phenology* literally means *the study of or a discourse on that which appears.* In another sense, though, *phaino* means “to let appear” and “to shed light upon.” While the difference may at first seem minimal, this secondary sense of *phaino* is vitally important to understanding how phenology prepares us to see ecological losses as worthy of our grief. Indeed, with this latter notion of *phaino* in mind, we arrive at a definition of phenology that goes something like this: *a study or discourse that sheds light upon, which lets appear.* Understood as a way of studying or discoursing, phenology has a revelatory power. It lets something be seen that we otherwise might not notice. Phenology does not just explain what *has* appeared; it also illuminates what is disappearing and what might yet be lost.

This revelatory power inheres in the activities that together comprise the work of phenology: witnessing, recording, and archiving. Witnessing orients us to the present, thus shedding light on both what appears and what fails to appear *now*—that is, at the moment of observation. For its part, recording orients us to the past. The written record, whether it takes the form of a note scrawled in a journal or an entry in an Excel spreadsheet or a post on social media, illuminates what appeared at a particular moment in time. And archiving orients us toward the future, toward what may continue to appear, and to what might yet disappear at a time to come. Phenological archives reveal both continuity and change; they disclose, in retrospect, what remains with us and what has been lost. The more robust these archives become—and there are good reasons to believe they are becoming more robust today—the more they shed light on those beings and ways of being threatened with erasure. This is so because phenological archives, when developed over extensive periods of time or at wide intervals, can reveal long-term trends. The more we understand such trends, the better equipped we are to anticipate—and perhaps stave off—losses to come.

Witnessing is at the core of phenology. All of the insights we gain from phenology are a result of someone witnessing something occurring in some place at some specific time. Although a few aspects of phenology can be carried out by instruments—for instance, the temperature can be recorded by a machine—there is not yet a substitute for living, breathing human beings who see, hear, smell, or otherwise sense what is going on around them. Phenology requires an attentiveness that, for now, only humans are able to furnish. In *Ground Truth*, Mark Hineline advises amateur phenologists to choose or make a phenological trail. "It need not be a straight line," Hineline explains.

> Zigzags are fine if they pass through varied landscapes and biomes. The best way to make sure of that variation is to change altitude, even if this involves only a few feet. The close-in part of your phenological trail should be a distance that you can traverse every day. Farther out, you might want to cover ground once or twice a week or, farther still, at least once a month.[48]

What matters, in other words, is that phenologists choose a path where they can be regularly present within the landscape. Although computers and books are the phenologists' friend, there is simply no replacement for the kind of fieldwork that Leopold and others pioneered. To notice what is going on in a land community, we have to leave the armchair behind and make physical contact with the more-than-human world around us.

While many people, as I will explain later, unknowingly contribute to phenological archives by recording and sharing their encounters with plants and animals, those who seek to bear witness to seasonal events often cultivate habits of attention that enable them to see and to hear more clearly. Of course, to a greater or lesser extent, we all perceive a range of phenological events, from the leaves changing colors in the fall to the circulation of pollen in the spring to the first snow in the winter. Unless we are deliberately paying attention to these cycles, however, we are unlikely to really notice the full scale of what is happening in our midst. As Leopold writes in "Prairie Birthday," "Tell me of what plant-birthday[s] a man takes notice, and I shall tell you a good deal about his vocation, his hobbies, his hay fever, and the general level of his ecological education."[49]

Phenology calls for something more intense than passive watching. Observation suggests a deliberate, active, responsive stance. The naturalist John Burroughs once wrote that, "The habit of observation is the habit of clear and decisive gazing; not by a first casual glance but by a steady deliberate aim of the eye are the rare and characteristic things discovered. You must look intently and hold your eye firmly to the spot, to see more than the rank and file of mankind."[50] To observe well, that is, one has to practice watching, listening, and learning. Phenologists never simply watch or listen; they also identify, to the best of their ability, what they are seeing or hearing, which usually requires some level of research to answer even the most basic questions: What have I seen or heard? What is the name of the phenological event I have witnessed? Correctly identifying and naming plants and animals takes more effort than many realize, as anyone who has tried to catalogue the plants on their property has likely discovered. Discerning what is going on with a particular species or within a specific ecosystem demands a good deal of attention and background research. My point is simply that the kind of witnessing in which phenologists are engaged is . . . well, engaged. It is an active attunement to what exceeds the human.

As we saw in the Leopold case, actively bearing witness to phenological events can draw us into more intimate relationships with our earthly cohabitants, whether they be the flowers in one's garden or the trees in nearby woods or the birds who migrate to and from one's region from year to year. This sense of intimacy and familiarity that bearing witness affords phenologists is a crucial precondition of an ethically oriented ecological grief. "We can be ethical," Leopold writes in "The Land Ethic," "only in relation to something we can see, feel, understand, love, or otherwise have faith in."[51] In *The Incidental Steward*, Akiko Busch echoes this sentiment. She speculates that, "at those times when we allow our esteem to be extended to those things we have made an effort to know well, a different kind of imagination comes into play, and I would suspect it is a manner of ardor that is, in the end, easier to sustain."[52] It is easier to love someone or something we have taken the time to get to know. Remember: knowledge begets sorrow. By really paying attention to what appears, and by actively seeking out knowledge of what we see and hear in our local environment, we expose ourselves not only to what is present but also to what might yet become absent. Everything to which the phenologist bears witness is already tinged with the possibility of loss; everything to which they attend may pass—or be pushed—out of existence. "Maybe," Busch recommends, "we bring a particular ingenuity to the process when these things could be lost so easily."[53]

In tandem with witnessing, recording is also at the core of phenology. In terms of preparing us to grieve beyond the human, though, it serves a different role. Where witnessing puts phenologists in touch with who or what appear *now*—and, by extension, who or what we are losing *now*—recording orients us toward the past. Written records of phenological events constitute a relatively durable trace of what was *once* present, what *once* appeared. Which is also to say: written records of seasonal events attest to a past state of relations. In 1945, red-winged blackbirds arrived on *this* day. In 1985, they arrived on *this* day. And in 2015, they arrived on *this* day. Notes that describe these events are the tangible traces of the way things were. Without records of this kind, it becomes difficult—if not impossible—to do anything other than speculate about the timing of seasonal events. The importance of written records, then, is that they make it possible to say, with a degree of certainty, that a particular phenological event has continued unchanged, been disrupted by climate change or more acute environmental transformations, or stopped happening at all. Written records help us talk about, and tell stories about, the way things were.[54]

Phenological records take a wide range of forms. Leopold, for instance, is rumored to have never left home without a small notebook in his pocket. In fact, when he perished in 1948 while assisting a neighbor put out a fire, he was carrying a notebook within which he had recorded his most recent phenological observations. Leopold documented, often in nearly illegible shorthand, the seasonal events he witnessed on a particular day. One day in April 1948, Leopold recorded seeing mallards, ducks, robins, hermit thrushes, and wood ducks. Alongside the names of the birds he saw,

Leopold often included information such as the time of day they were seen, how many he observed, the intensity of the light when he witnessed them, the temperature, and cloud cover. Leopold later entered these informal notes into his "Shack Journals," running accounts of both the human and more-than-human goings-on at the family's farm on the Wisconsin River. From 1935 until he died, his records accumulated. In his final years, Curt Meine observes, "His journals grew heavier than ever with the comings, goings, callings, feedings, leavings, and flowerings of more species than ever before."[55] Here, taken from the 1946 "Shack Journal," is a representative phenology report:

> Saw first marsh hawk flying south over marsh. Could not tell sex. Saw a monarch caterpillar (black with a brown band around—steps on ground in front of shack). No pheasants crowed. No cardinals sang. Lewis says he saw a robin in his yard "about 2 days ago,"—saw none this winter.[56]

As these informal records accumulated in the "Shack Journals," Leopold combined his data into phenological tables. Along the left side of the tables, he listed phenological events. On the top, he listed the years. As time passed, he populated the fields with the exact date when he witnessed each event take place for the first time. These tables present a time-lapse view of the year-to-year rhythms of earthly life in central Wisconsin at particular moments in time.

Like witnessing, keeping records of seasonal events draws phenologists into more intimate relationships with the more-than-human beings and ways of being with whom we cohabit the earth. This sort of intimacy is the consequence of curiosity entwined with, and perhaps animated by, care. "Caring," writes Donna Haraway, "means becoming subject to the unsettling obligation of curiosity, which requires knowing more at the end of the day than at the beginning."[57] In particular, recording what one has seen or heard requires that one "take care" with the facts. Written phenological records need to be accurate if they are to usefully teach us about the past, if they are to shed light on what remains and what has withered away. In *Walden Warming*, the biologist Richard Primack explains that, "the first step is to learn to identify the trees, shrubs, and wildflowers, the birds and insects, and other creatures that live in your area."[58] Which is to say: the practice of recording observations demands that one develop at least some rudimentary knowledge about the animals, plants, and other entities one hopes to document. Committing one's observations to writing invites a degree of carefulness that simply witnessing does not. This is so, at least in part, because the written record is meant to testify to something that has happened in the past. It makes little sense to keep sloppy records; in fact, no record at all might be preferable to carelessly produced ones. When we produce accurate records, we are effectively writing history in an attempt to "make some sense out of a changing landscape."[59] And while it is surely possible that someone might be willing, or even eager, to distort the phenological record for some reason, most people who spend their time observing the natural world are committed to getting natural history correct because, to put it plainly, the future of earthly coexistence depends upon it. Indeed, if we wish to make wise decisions in and for the future, we need to understand the past.

If witnessing sheds light on current losses and recording illuminates past losses, the archives that phenologists generate reveal losses to come. This is especially true in a period of earth history characterized by climate change. Against the horizon of wide-ranging and irreversible ecological and earthly transformations, change and loss are inevitable. It is for this reason, perhaps, that people are now turning—and in great numbers—to the practice of phenology, which is itself a fundamentally archival undertaking. "Like the discipline of academic history," Hineline explains, phenology "is grounded by a textual record—tens of thousands of observations, recorded with accompanying notes about time and place."[60] It is grounded, in other words, by archives. In *Archive Fever,* Jacques Derrida argues that the inevitability of loss sets our archival desire aflame: in the face of losses to come, we attempt to safeguard even a trace—indeed, often only a textual trace—of what might yet be lost. "There would indeed be no archive desire," Derrida writes, "without the radical finitude, without the possibility of a forgetfulness which does not limit itself to repression."[61] To stave off the loss of both matter and memory, we archive. We gather together. We set aside. We hold. We conserve. We save. And so, we might well regard phenology in the twenty-first century as—to quote Ann Cvetkovich—an "urgent effort toward preservation in order to grapple with loss."[62]

What, though, is an archive? And in what sense can phenology be said to constitute an archival practice? Archives have traditionally been conceived as repositories of the past, physical locations—often housed within broader institutions—where material artifacts are held, studied, revered, and preserved. So conceived, archives furnish historians and others interested in telling the story of the past with the raw materials—books, letters, handwritten notes, photographs, artworks, clothing, objects, and much else besides—from which stories are spun. Itself ancient sounding, the word *archive* conjures images of gray-haired scholars wearing bright white gloves and wielding magnifying glasses as they flip through the brittle pages of ancient, and therefore important, manuscripts.

And what does this scholar hope to achieve in the archive? They hope to *discover* something about the past—namely, what *really* happened. From this angle, archives appear to be the scene of a discovery, the place where one goes to uncover the truth of the matter. At first glance, this seems to be the case with phenological archives, too, since they provide an account of past seasonal, natural rhythms. If one wants to know what once happened, all one has to do is enter into these archives and trace the unfolding of cyclical events in the myriad records left behind by people like Leopold and Thoreau. And, to a certain extent, this is true: the archives that phenologists generate possess a referential power. Insofar as they serve the needs of science especially, verisimilitude is a vital quality of phenological records: if they are to shed light on the past, the archives need to contain information that is accurate.

However, archives cannot—or should not—be reduced to their referential function. Alongside scholars in the field of archival studies,[63] rhetorical scholars have been working to trouble a seemingly straightforward understanding of archives. "Archives are not simply repositories of information about [the] past," Pamela VanHaitsma explains in a discussion of this research, "but sites of ongoing deliberation and invention."[64] Rather than treating archives as stable scenes of discovery in which the truth of the past is waiting to be found and decoded, rhetorical scholars such as Chuck Morris, Ekaterina Haskins, and Barbara Biesecker invite us to understand archives as sites of argumentation and creation. Archives, Biesecker claims, cannot "*authenticate* absolutely" the truth of an irretrievable past so much as they "can (be made to) *authorize*" particular, situated, interested interpretations of events which have already occurred. So, yes, archives provide the materials—Biesecker calls them "the elements of rhetoric"—from which a story is constructed.[65] But, crucially, it is not a single, definitive, undisputable story that the archive authorizes. Rather, as Chuck Morris pithily puts it, the archive is, or can become, an "inventional wellspring," a source from which a collection of stories—some hegemonic, others resistive—find nourishment.[66] Archives are, as it were, put to work in the service of very specific stories about the past, which means that some elements are highlighted while others are obscured.[67]

Archives can also be put to work in the imagination of futures, and this is especially so in the case of phenological archives. The future Leopold imagined was marked by more of the same; phenological archives, he hoped, would disclose enduring earthly rhythms and make possible an understanding of the delicate balance achieved by the ongoing interplay of animals and plants, habitats and ecosystems, weather and climate. And yet, as Hineline observes, "When a phenological record is made and kept, whether it was intended for such a purpose or not, it may become useful for new sorts of analysis in time."[68] At the moment one contributes a record to the phenological archive, one cannot predict what it will reveal in the decades and centuries to come. As we saw in the first part of this chapter, Leopold's records have been mobilized to tell a story of transformation and loss in central Wisconsin, a story about the unmaking of ancient patternings, the erasure of beings and ways of being alike. Such stories make clear the urgency of action in the present. Drawing on Leopold's records from the 1930s and 1940s, his followers shed light on a history of quiet and nearly invisible losses. The gloomy undertone of such stories is that those losses continue today; inconspicuously, beings and ways of being are unraveling under the pressure of climatic changes. Contemporary phenologists are creating the archives from which the story of this future—a future characterized by the slow and steady disappearance of familiar plants and animals, as well as the seasonal rhythms as *Homo sapiens* have long known them—will be constructed.

It is in this sense that phenological archives draw our attention to future losses. By revealing past and ongoing transformations and losses, these archives enable us to fashion well-grounded narratives about all that we stand to lose. Narratives about the way things might be in the future imply a sense of who "we" will be in the future, too. As Biesecker puts it, "Whatever else the archive may be—say, an historical space, a political space, or a sacred space; a site of preservation, interpretation, or commemoration—it is always already the provisionally settled scene of our collective invention, of our collective invention of us and it."[69] What version of ourselves do phenological archives suggest, then? In the last part of this chapter, I hope to show that these archives, always in flux and always in the making, are one of the conditions of possibility for an ecological mourning to come. That impulse to witness, to record, and to archive at the heart of phenology itself enacts the work of mourning, but it also prepares a *people to come* to grieve *losses to come*, to awaken a future "us" to an emergent "world of wounds," so that this future "us" can acknowledge and grieve those losses as yet on the horizon but also so that "we," today and tomorrow, might yet work to conserve, to safeguard, and to tend those who find themselves under threat of erasure.

Revealing Losses to Come

What I have been referring to as "the phenological archive" is, in practice, nothing singular. It is perhaps more appropriate to write of phenological *archives,* for the records comprising phenology are themselves dispersed across an array of sites—some physical, some digital; some systematically structured, some informally organized; some preserved by venerable institutions, some maintained by vernacular associations. Phenological archives may exist in personal, private spaces—in a desk drawer, for instance, or within an Excel spreadsheet hidden deep within someone's computer hard drive. Or they may exist in communal, public spaces. As we will see, phenological archives sometimes emerge in unexpected places: on Facebook and Instagram, for example. And sometimes phenological archives migrate—or are moved—from private to public domains, as was the case with Leopold's handwritten records scribbled in notebooks and organized into journals. Today, one can scroll through Leopold's phenological records and tables online as well as in person at the University of Wisconsin's special collections library in Madison.[70] My point is simply that phenological archives assume a variety of forms, forms which are responsive to the conditions under which they are created.

The intensification of climate change and the spread of the internet have given rise to a recent renaissance in amateur phenology. Phenology is one among many practices that so-called "citizen scientists" are taking up with renewed vigor.[71] As Akiko Busch writes in her book, *The Incidental Steward,* "the search for some sense of belonging seems to befall many of us, whether it is in watching, naming, counting, documenting, or otherwise noting the natural events unfolding around us."[72] The specter of climate change has imbued these practices with a sense of significance and urgency. This is so because many cyclical events are both responsive to and compromised by climatic changes. The changes are wide-ranging and interconnected. In *Walden Warming,* Richard Primack illustrates the ways in which climate change sets off cascades of overlapping transformations:

> What might happen if global warming messes with the timing in our neat little food web? The tree now leafs out earlier than it used to, and the caterpillar matches the change, as they are both responding to the same

> *local* changes in temperature. But our bird misses the boat—from Ecuador, it can't tell whether it will be a warm spring or a cold spring in its northern breeding territories. It continues to migrate and breed at the same time each year, using length of daylight as its cue for departure for the trip north. When it finally arrives, the bird breeds too late to feed the caterpillars to its chicks. The oriole chicks could die of starvation unless the parent bird can feed them something else. Let's suppose our particular bird is flexible and learns to eat a multitude of other insects that happen to be around when it breeds. One of these unfortunate insects is our bee. The bee population declines dramatically because the bird is now eating it, whereas before it was mostly free from predation. As a result, there are not enough bees to pollinate the violet flowers. Consequently, the violets don't produce any seeds, and the population begins to decline. Meanwhile, the caterpillar is doing splendidly, eating lots of tree leaves.[73]

As this example suggests, tracking the timing of seasonal events alongside the acute effects of climate change sheds light on the ways in which processes that are planetary in scope nevertheless insinuate themselves into the warp and woof of particular places.

If the coexistential threats posed by climate change have made phenology feel more urgent at the beginning of the twenty-first century, the internet—and the myriad devices and platforms it gives rise to and coordinates, from smartphone cameras to phenology apps to virtual social networks to collaborative data gathering websites—has shaped *how* many people go about witnessing, recording, and archiving seasonal events. We cannot separate archival technologies from the act and experience of archiving. As Derrida contends, "the technical structure of the *archiving* archive also determines the structure of the *archivable* content even in its very coming into existence and in its relationship to the future." Or more to the point: "The archivization produces as much as it records the event."[74] Even as the internet and its entailments seem to render time and place irrelevant, these technologies also orient us to the earth in new ways. In particular, the internet makes possible two ideas. First: anything that can be saved should be saved. And, second: public participation in the archiving process will lead to more diverse, and therefore more robust, archives.[75]

A few examples help to illustrate these points. Concerted efforts to collect and keep phenological records such as the USA National Phenology Network exploit the internet's vast storage capacity: more than fifteen thousand participants have already uploaded more than fifteen million records via the network's Nature's Notebook program, which now includes an application for iPhones and Androids.[76] The records they submitted are entered into a database that program participants, scientists, and others can download and use for their own investigations. Informal phenological networks, such as the Duluth Phenology Facebook group, also invite participants to share their observations—often accompanied by photographs—of the goings-on of plants and animals in particular locations. Although these posts are not typically systematized in any way, over time they may become useful to those with an interest in understanding how seasonal rhythms have shifted.[77] But people need not intend to create phenological records in order to add to phenological archives. Instagram users have, for example, tagged more than two thousand photos with the hashtag #amanitamuscari, the scientific name of the iconic toadstool mushroom. And more than a half million images bear the hashtag #bluebells, the common name for the delightfully dangly *Hyacinthoides non-scripta*. Without even realizing it, Instagram users are passively contributing to phenological archives that are swelling to new proportions in the digital era. As I have written elsewhere,

> Ongoing and collaborative phenological work, predicated though it may be on the smartphones and social media that so often pull us away from the earthly, gives me hope, for it is only through understanding the plants and animals with whom we cohabit the earth that we stand any chance of appreciating what might be lost in the ongoing planetary crises—crises, it must be said, which make the work of earthly coexistence our most pressing and persistent task.[78]

When many people in many places at many times turn their attention to the land's rhythms, we collectively become more capable of "appreciating what might be lost." Phenology has often been undertaken as a communal, networked practice. As Richard Hopp explained in 1974, phenological observation networks comprise "cooperators observing and recording the date of the occurrence of defined developmental stages or phenophases of indicator plants."[79] In the 1930s and 1940s, for instance, Leopold and Jones relied on at least nineteen informants in Sauk and Dane counties to supplement their own records.[80] From the mid-1950s to the early 1990s, Joseph Caprio—then a professor at Montana State University—orchestrated a still more extensive phenological network spanning twelve western states in which contributors focused their attention on a single widespread plant, the common lilac (*Syringa vulgaris*).[81] From Caprio's efforts emerged the USA National Phenology Network of more than fifteen thousand volunteers and records of more than fifteen million individual phenological events.

How, though, do these phenological records come into existence? At the core of the USA-NPN is the Nature's Notebook program. When volunteers sign up to participate in the "online plant and animal phenology monitoring program," they receive basic training on how to observe, record, and submit their observations.[82] New participants are prompted to create a site profile or, in other words, to define the parameters of the location(s) where they will conduct routine observations. Often the site is simply the participant's backyard, a nearby park, or a path they walk often. Participants are also prompted to select the species of plants, birds, mammals, insects, fish, and reptiles they will observe at their site. One can select as many or as few species as one feels comfortable observing; over time, participants can also add new species to their lists. As the ecological critic Scott Slovic once remarked, "The verbalization of observations and reactions makes one much more acutely aware than would a passive assimilation of experience."[83] Hence, I suspect many Nature's Notebook participants begin to actually see more species as they develop their record-keeping skills. Regardless of whether one's list grows over time, the program generates customized phenology tables that can be filled in each time participants make observations. Once the tables have been completed for a given period (usually several weeks), participants manually record their data through the program's Observation Deck. Alternatively, participants can enter observation data using the USA-NPN's smartphone app, also called Nature's Notebook. Both methods for recording observations—pencil-and-paper and smartphone app—prompt participants to look for the same species each time they conduct their fieldwork, and each option has built-in explanations (and sometimes images) of the phenophases they should take care to watch out for.[84] In case participants forget to record their observations, the network deposits a gentle reminder in their email inbox each week.

As participants submit their observations to Nature's Notebook, they join millions of other records in the USA-NPN's database. As records accumulate over many years, they shed more and more light on persistence and loss in specific ecological systems. These data are especially useful to ecologists and other scientists for whom phenological records can yield new insights about the acute, local impacts of climate change. The network provides its data free-of-charge to researchers, who have already published more than forty scientific articles based on participants' records.[85] These recent observations can be compared to data collected by earlier phenologists to determine changes over a given period time. Drawing upon records made throughout a one-hundred-and-eleven-year period, for instance, a team of researchers found that both cutleaf toothwort (*Cardamine concatenata*) and yellow trout lily (*Erythronium americanum*) in West Virginia now bloom earlier than they did a century ago. Both of these spring ephemerals, which are especially sensitive to warm temperatures, advanced their spring flowering by six days. Earlier blooming, the researchers concluded, puts both species "at risk for phenological mismatches."[86] *Cardamine concatenata* may be at a particular reproductive disadvantage as a result; as spring arrives earlier, the pollinators it depends upon—bees—may not be available. Hence a "phenological mismatch." Because the blooming dates advanced less rapidly at higher elevations where temperatures remained lower, it may also be the case that specimens at lower elevations struggle to survive at all in the future. It is only when records of blooming dates and other phenophases are present over extensive periods of time that such revelations occur. Phenological records, Primack notes, "increase in value with the addition of each new year."[87]

It is not only scientists who find value, interest, and meaning in accumulations of phenological records. The USA-NPN deploys its participants' observations to generate its "Status of Spring" update and visualization each year. In essence, the "Status of Spring" is an annual report on how early or late the onset of spring occurred in specific locations. The yearly report is based on "observations of the timing of first leaf and first bloom in a cloned lilac cultivar . . . and two cloned honeysuckle cultivars," both of which "are among the first woody plants to leaf out and bloom in the springtime and are common across much of the country."[88] Using these data, the network creates a series of animated maps illustrating the onset of spring throughout the mainland United States. As the date in the bottom, left corner of the map progresses from January 1 to the present date, shades of red and blue slowly and then quickly spread over the states. Lighter shades represent a typical spring onset whereas darker shades indicate a significant departure from usual spring onset. The darkest shade of red, for instance, indicates that spring arrived twenty days early in that area. The darkest shade of blue signifies that spring was twenty days late. The map for 2020 included massive swaths of dark red in the south, northeast, and west, as well as significant regions of dark blue at the northern latitudes from Maine to Montana. In short, huge parts of the country experienced spring arrivals that were nearly three weeks early or three weeks late. This manner of presentation tends to highlight deviations from a specific yet vague "norm."

In delineating such deviations, the USA-NPN's "Status of Spring" maps do not so much confirm ecological losses as they imply them. Indeed, although neither the maps nor the accompanying text explicitly calls attention to losses *per se*, they nevertheless insinuate that year-to-year disturbances are now the norm. Abnormality has, in other words, become normal. Their rhetorical power lies in their suggestiveness. Change, not continuity, is to be expected. Loss, not persistence, can be presumed. By indicating disorder, these maps draw attention to the relationship between climate change and the seasons: what we could once take for granted is no longer certain. They register something uncanny about what it means to dwell on the earth today. For instance, the release of the 2020 "Status of Spring" maps prompted a discussion on *The Weather Channel*. In a short segment, USA-NPN's associate director Theresa Crimmins talked with hosts Stephanie Abrams and Jim Cantore about how to read the maps. At one point, toward the end of their conversation, Cantore asked aloud: "Can't we just have a normal season?"[89] The two broadcasters cast an uneasy smile at the cameras, thank Crimmins, and then conclude the segment. I gesture to this moment because it highlights the ways in which phenological archives reveal to us how different the future will look from the past. In some significant sense, Cantore's apprehension of change and disruption in the present—the fact that seasons are no longer "normal" as well as the fact that what used to be considered "normal" no longer exists—gestures toward the peculiar kinds of loss that phenological research is forecasting. What is at stake are not only specific plants and animals or particular species but, rather, *ways of being*.

Cantore's question—"Can't we just have a normal season?"—points at once to the past and the future. The question presupposes that what happened in the past was "normal." Scientists describe this as a "baseline" understanding. Individuals tend, the theory goes, to presume that the conditions under which they matured are typical, natural, and "correct." For instance, if you grew up ice skating on a particular pond in mid-February and then, later in life, came back only to discover that the pond is ice-free most of the winter, it is likely that you would interpret the recent state of affairs as a deviation from the usual and, thus, as a loss of both normalcy and stability. Cantore's question, provoked by the USA-NPN "Status of Spring" maps, registers this sense of loss and also projects it into an uncertain future.

His question subtly suggests its own answer. He does not ask "when" the seasons will return to normal. Rather, he exasperatedly asks, "Can't we just have a normal season?" His question suggests that the seasons are already so out-of-kilter that there is likely no turning back. *Can't we* already seems to imply *we can't.*

Thus, the strange headspace we now occupy: we recognize that humans' collective actions in the past and the present permanently alter future possibilities, although we often struggle to comprehend what futures our behaviors curtail or make less possible. We know that the seasons will never be normal again, if by "normal" we mean the way things were before the "Great Acceleration" (beginning in the middle of the twentieth century) or some other baseline. But many among us also have a hard time accepting what we have already lost and what else we will lose in the future. Phenological archives shed light on continuity and change in the past. In doing so, they show us the profoundly contingent and precarious quality of every baseline: what we take for granted today may not exist tomorrow.[90] The compulsion to witness, to record, and to archive what appears today—the compulsion at the heart of phenology—is a means of preparing ourselves to recognize and to grieve losses to come. Phenology, in the end, is always "an urgent effort toward preservation in order to grapple with loss."[91] The archives that amass cannot save the material relations to which they attest, of course. But they can and do preserve a memory of the past. And it is in contrast with that archival memory that we become capable of assessing the present and imagining the future. Indeed, in phenological archives we perceive not just the endurance of ancient rhythms but also the unwinding of particular ways, human and more-than-human alike, of dwelling on earth.

CHAPTER 4

Imagining Loss

On Making Visible

Imagining *Tsuga canadensis*

Driving south out of town in early August, I watch the world go by through the passenger's side window as my partner steers our hatchback along unfamiliar roads, around unexpected curves, through novel territory. Aging shopping centers and new condominiums give way to used car lots and roadside produce stands, which in turn give way to farmland, red barns and pastured horses and bales of hay all visible as we speed by. Having reached Pine Grove Mills, population 1,500, Brandon swings a left onto South Water Street, thus beginning our fossil-fueled ascent of Tussey Mountain and our entrance into Rothrock State Forest, nearly 100,000 acres of protected forest land encompassing our destination: Alan Seeger Natural Area.

To reach Alan Seeger, we coast down the other side of Tussey before leveling off into rural countryside. Rolling hills and dense forests form the horizon on either side of the two-lane highway. Although no street signs indicate our way, we know—thanks to Google Maps—that an Alan Seeger Road is up ahead on the left. As it turns out, the road *is* marked: a patch of carefully tended flowers and a TRUMP 2020 campaign flag greet us as we swing abruptly onto Route 1023, a still narrower blacktop that leads us past quaint farmhouses, two churches, a rural cemetery, a sheep farm, a dairy farm, a Christmas tree farm, and much else besides. The lane further narrows as we enter the dense woodland which drew us here. What scant sunlight pierces the canopy above dapples the forest floor as we leisurely make our way to a gravel parking lot.

The owners of the only other car in the parking lot depart as soon we arrive, leaving us alone, together, among the white oaks (*Quercus alba*), red maples (*Acer rubrum*), white pines (*Pinus strobus*), and American rhododendrons (*Rhododendron maximum*) that make their home here. But it is a different sylvan species that summons our attention as we amble along the mile-long Alan Seeger Trail, which cuts a loop through fragments of old-growth forest. Spared "due to a boundary-line dispute between competing logging companies," the eastern hemlocks (*Tsuga canadensis*) we have come to see reach heights of more than one hundred feet; their radial boughs and dense, lacelike foliage create a woodland environment that is characteristically dark, hushed, and, to we contemplative *Homo sapiens*, meditative.[1] "When we enter a hemlock forest on a sunny day," writes David Orwig, "we sense a change and immediately recognize that we are in a different and special place."[2] Indeed.

One of the most charming parts of walking the Alan Seeger Trail is weaving one's way, childlike, through the giant rhododendrons which arch over the path to form a verdant tunnel. Backs bent forward, eyes level ahead, we crouch through the tangled trees until we once again can stand and observe our surroundings. Quite by accident, we notice a rustic timber sign, into which is carved the following: "EASTERN HEMLOCK / TSUGA CANADENSIS / OVER 500 YRS OLD." Intrigued, we look around expecting to spot the specimen to which the sign refers. Though we see a few hemlocks over the rhododendrons, none appears tall or girthy enough to be a half-millennium in age. Not far from the sign, a downed tree, soft with age and covered by moss, lays to one side of the trail. We speculate that perhaps this is the ancient tree, the sylvan elder, and so begin to view the sign in a new light—not, as we at first had, as a marker of something present but, rather, as something of a tombstone, an emblem of absence.

Conjuring Loss

Much like a tombstone in a cemetery, the wooden sign along the Alan Seeger Trail signifies a loss. Tucked into the rhododendron, the unassuming marker makes the eastern hemlock's absence conspicuous: noticeable, palpable, disconcerting. It is a sign without a referent; its object evades it—and us. It gestures toward a nothingness, leaving us to imagine what this ancient hemlock was like, how big around it was, how high it reached, what birds and other woodland creatures it harbored, what rains and droughts it endured, what winter storms it withstood, what history it witnessed. Right or wrong, I choose to imagine that the decaying stump not far from the sign is the foundation of the tree in question—three or more feet in diameter, it seems a reasonable candidate. In my mind's eye, I picture the absent hemlock biding its time in the understory until one of its elders collapsed, enabling sunlight to stream onto the forest floor for the first time in ages. Patient tree that it was, our eastern hemlock seized its opportunity, devoured the sun's rays, grew new rings of wood like never before, inched toward the sky, and replaced the elder whose life and death its own existence surely presumed. The mere presence of the sign evokes an imagined past, an imagined history, an imagined life.

Of course, as I have been suggesting, the wooden marker also enables passersby to imagine an ecological loss that would probably remain inconspicuous if it were unmarked. Placed along the trail when the ancient hemlock still stood tall, the sign does not so much announce a loss as it generates a structure of anticipation. Although the tombstones we find in cemeteries usually disclose birth and death dates, the sign divulges an approximate age: "OVER 500 YRS OLD." There is no indication, at least no verbal indication, that the more-than-500-year-old eastern hemlock has died—no parentheses, no dates separated by a dash somehow meant to convey the time of a life. Thus does the sign invite hikers to search for something that is no longer there, someone whose form has shifted, whose presence is less perceptible than it once was. Looking up from the sign, one expects to see a hulking sylvan body towering above the understory. Finding none, it becomes apparent that one occupies not an unchanging nature but a dynamic land community, one where nothing stays the same forever. Loss, death, and decay, though pervasive in forests as in other land communities, hardly call attention to themselves, overshadowed as they are by the abundant presence of life's myriad modes and forms. In ways its creators may or may not have foreseen, the sign makes a wound visible, enticing those who notice it amid the rhododendron's waxy green leaves to imagine the life and loss of a "forest giant."[3]

Students of rhetoric have, since the discipline's birth in ancient Greece and Rome, acknowledged the power of imagination. When rhetorical scholars discuss the imagination, their remarks are often embedded within broader considerations of the emotions. Aristotle, for instance, takes up the relation between fear and imagination in Book II of his treatise on *Rhetoric*. "Fear," he explains, "may be defined as a pain or disturbance due to imagining some destructive or painful evil in the future."[4] We fear that which we imagine will cause us harm. The feeling of fear, then, far from being a simple reaction to an external stimulus, arises from a structure of anticipation that depends on our capacity to conceive of a future in which something painful comes to pass. Though Aristotle's insights about the emotions are interesting in their own right, they serve in his *Rhetoric* to illuminate how orators can "put [their] hearers, who are to decide, into the right frame of mind." Effective speakers, in other words, understand the emotions—what they are, who experiences them, and what provokes each one—and are thus able to stimulate feelings in their listeners, feelings which ultimately influence their judgment. As Aristotle put it, "When people are feeling friendly and placable, they think one sort of thing; when they are feeling angry or hostile, they think either something totally different or the same thing with a different intensity."[5] If an orator wishes to evoke fear among their listeners, they would do well to spark in them an ability to imagine that their well-being is threatened by some forthcoming event. Having so stoked their imagination, Aristotle claims, the orator's audience will make decisions in accordance with their newfound sense of fear. In short, rhetoric begets imagination; imagination stirs feelings; feelings drive judgments.

Traditionally conceived, rhetoric's *telos*—its purpose—is judgment. We talk to and with one another with the aim of swaying, subtly and not so subtly, one another's decisions; indeed, the optimism that animates rhetorical practice emerges from the conviction that we can transform the world by changing how people feel, see, think, and act. For Aristotle, the threefold nature of temporality—its division into past, present, and future—occasions three distinct yet interrelated kinds of rhetoric: forensic, epideictic, and deliberative. These three species of rhetoric are born out of the need to make decisions about events that have already occurred, are presently unfolding, or might come to pass.[6] Rhetors who want to enable audiences to render judgments are, then, faced with the task of crafting performances and texts that stimulate their interlocutors' ability for imagining what is not immediately before them. "Rhetoric's work," Michele Kennerly explains,

> often consists of giving presence to the *unseeable*—something not yet or never capable of being seen—or to the *unseen*—something visible but ignored. A demonstrated need for judgment on matters past, passing, or to come gives impetus to bringing the absent or unapparent into view. To show something or someone previously unseen due to an expanse of distance or time, or previously unrecognizable due to a deficit of attention or feeling, a rhetor must appeal to the same mental faculty that enables self-indulgent escapes into fantasy.[7]

To put it differently, rhetoric's work often consists of conjuring a sense of what remains invisible or occluded, hidden from view and, therefore, excluded in certain significant ways from the realm of what matters, what captures our concern, what beckons our care.

One way to view rhetorical practices, then, is to understand them as so many ways of making things matter, of giving them salience by bringing them to light in interesting or new ways.[8] The rhetorical means for accomplishing salience may be more or less complex, more or less sophisticated. As this chapter's opening vignette demonstrates, an unassuming sign along a walking path goes a long way toward generating an environment within which the absence of a singular tree comes to matter to passersby like me and my partner. It is not that the sign produces a stable meaning; it does not, after all, even point to the something it names and describes. Its rhetorical force has less to do with meaning and more to do with how it sets an adventure into motion as walkers seek out the ancient tree it gestures toward. Discovering no physical entity that fits the sign's bill, we are forced to imagine—to create a mental picture of—the absent tree. Or, at the very least, we are invited to embark on such an adventure. When we are (or let ourselves be) pulled along by the sign's gestural rhetoric, we find that the missing hemlock, as well as the forest around us, concern us in ways they had not before. Having brushed up against the possibility of loss, having conjured the dead *Tsuga* in our minds, we begin to see this grove of old-growth trees not only as the resilient land community that it is but also as precarious, lose-able.

Making (Ecological Loss) Visible

I am reminded of another, more spectacular effort to summon attention to the fragility of a singular tree and, by extension, the temperate rainforest where it put down its roots. While the sign inadvertently gestures to the absence of its past referent, when Julia "Butterfly" Hill took up residence in a thousand-year-old redwood tree in northern California, her bodily presence in the 200-foot-tall tree signaled its imminent demise. The tree, which Hill and other Earth First! activists named "Luna," was to be felled by the Pacific Lumber Company. Hoping to block the loggers from killing Luna, Hill lived for more than two years—738 days, to be exact—on a platform perched 180 feet up in the tree's canopy. Hill's tactic proved successful. In the end, Pacific Lumber eventually backed down, agreeing neither to fell Luna nor any of the trees within a 200-foot radius. In *Image Politics*, Kevin Michael DeLuca explains how Hill's occupation of Luna transformed the tree from a potential commodity into a singular sylvan being worthy of care and concern. "Through inhabitation," DeLuca notes, "a space that is interchangeable with many others becomes *this* place that is irreplaceable in its particularity."[9] By dwelling in Luna's canopy, in other words, Hill's precarious presence heightened onlookers' capacity to see Luna itself as precarious. By placing her body at risk, Hill opened spectator's eyes to the very material ways in which commercial logging puts redwoods' continued existence at risk.

Hill's spectacular gesture is one among many examples of environmental activism that exploit the rhetorical potency of the visual field to render ecological losses, or potential losses, more real, more tangible, more pressing. Consider Eliot Porter's *The Place No One Knew: Glen Canyon on the Colorado*, a book of large-format photographs of Glen Canyon in the years just before it was flooded to form Powell Reservoir. Published in 1963, Porter's collection captures scenes from the canyon's prediluvian life. The book was a belated, last-ditched effort to halt construction of the dam, an effort that, although it failed its immediate purpose, now makes it possible to imagine and to measure what was lost when the gates of the dam were opened at last.[10] Consider, too, Jeff Orlowski's documentary *Chasing Ice*. The film follows James Balog's efforts to document the rapid melting of glaciers in Greenland, Iceland, and Alaska using time-lapse photography. Not only does the film lay bare the loss of the glaciers; it also reveals one of the processes through which ongoing ecological losses are made visible and tangible.[11] Or, to take just one other example, consider the Audubon Society's Survival by Degrees project, a publicly accessible data visualization tool that enables users to examine habitat loss for endangered avian species in North America under three warming scenarios. Toggling among 1.5°C, 2°C, and 3°C global average temperature increases, users are invited to imagine how various degrees of climate change, and various degrees of climate mitigation, will shape the future distribution of birds in Mexico, Canada, and the United States.[12]

Photographs, films, maps—all are familiar objects for the student of visual rhetoric, for whom, obviously, language is not the only show in town. Although art historians have always taken the visual seriously, "the embrace of visuality in the humanities would happen more gradually."[13] An array of intellectual, political, and technological transformations in the middle of the twentieth century led, at long last, to the field of communication's tenuous embrace of rhetoric's visual modes and forms.[14] My aim here is not to recount that history; others have already produced highly useful histories of the discipline's movement beyond oration and writing to include diverse kinds of artifacts and performances, including those that are primarily visual in nature.[15] In line with this book's larger purpose, my aim is rather to underscore how ecological losses past, present, and to come are *made visible* and, thus, conceivable, tangible, imaginable. The artifacts I study in the next several sections of this chapter—a memorial to the extinct passenger pigeon (*Ectopistes migratorius*), site-specific installations of Greenland's melting glacial ice, and a set of speculative images depicting imagined climate change futures—demonstrate, each in their own way, the force with which visual rhetorics set the stage for feelings of ecological grief by shifting whether and how audiences see other animals and plants, species and land communities, ecosystems, and ecocultural assemblages as lose-able.

As the examples mentioned above make clear, there are many ways to make visible what has previously gone, or usually remains, unseen. Far from being neutral or inevitable, the visual field is organized and inflected so as to render some dimensions of reality visible while leaving others occluded and, thus, harder to see. Representation matters, not because visibility leads necessarily to liberation but, rather, because what appears in public becomes common. Shared ideas and feelings and responsibilities flow forth from what we have in common, from what we all are capable of seeing.

It matters, for example, whether images of people who dwell in toxically assaulted communities—places where polluting industries sacrifice the health of humans and other members of the land community for profits—circulate publicly. When we are able to see the human and ecological costs of reckless industrial practices, when we are forced to look into the eyes of those whose lives could be lost, we become responsible for and to those beings whose existence our material comfort puts at risk.[16] So long as those whose lives are at stake remain marginalized, both spatially and rhetorically, it remains difficult for wider swaths of the public to see how our own lives are often violently interconnected with others. It matters, too, whether images of anthropogenic ecological destruction appear in public, whether the ugly byproducts of our fossil-fueled, resource-intensive ways of life are laid bare for all to see. Representations of ecological degradation are often complicated, as Jennifer Peeples notes, by the fact that tailing ponds and rock quarries and oil spills can, from at least some angles, be awfully beautiful, even awe-inspiring.[17] In the end, then, such images reveal not only what has been or could be lost—"pristine" landscapes—but also an ambivalence at the heart of modern life, a sense of disgust tempered by a sacrificial logic that sanctions ecological destruction so long as it can be confined to an elsewhere.[18]

I borrow the phrase "making visible" from Davi Johnson, who deploys the term to describe Martin Luther King Jr.'s task *and* tactic during the 1963 Birmingham Campaign. Johnson explains that, by the early 1960s, the once spectacular displays of violence against black bodies, particularly in the form of public lynchings, had largely receded into private arenas and, thus, had become less visible than they once were. King's task, Johnson notes, was to "[make] the reality of racism immediately visible to an audience of white moderates content in their complacent avoidance of overt conflict."[19] As King's followers staged nonviolent demonstrations in Birmingham, Alabama, they were aware that their actions would likely attract a violent reaction from the city's racist commissioner of public safety, one Eugene "Bull" Connor. When, under Connor's orders, police officers turned fire hoses and snarling dogs on the nonviolent protestors, photographers and reporters were waiting to report the story. "For King," Johnson contends, "the best means of making racism visible was by exposing its action on black bodies."[20] King's tactic was effectively to bring racism back into the light of day, to expose the violence which, for myriad reasons, had begun to slip out of public view and so out of public consciousness.[21] Not only did the photographs of the encounters between protestors and police literally make violence visible; they also crucially enabled viewers to see and to imagine racism's ongoing legacy. By bringing racism before audiences' eyes, the Birmingham Campaign made it possible to see something that was hiding in plain sight, something present yet difficult for most to see. To make visible is both to represent and to disclose, both to depict and to reveal, and it is in this doubled sense that I use the phrase "make visible."

Like naming what we might lose and archiving ecological continuities and changes, making visible ecological losses past, present, and to come is a path to ecological grief. As I have written elsewhere, "it seems to me that we desperately need new ways of seeing that enable us to apprehend the (un) grievability and precarity of [the] more-than-human beings, species, places, [and] ecosystems" with whom we share the earth.[22] In the following pages, I visit the Monument to the Passenger Pigeon in Prairie du Chien, Wisconsin; reflect upon *Ice Watch*, artist Olafur Eliasson's installation of Greenland's melting glacial ice in Paris, London, and Copenhagen; and, finally, train my vision on digital artists Robert Graves and Didier Madoc-Jones's "Postcards from the Future" project. What I hope to show with these examples are some of the ways in which visual rhetoric exposes audiences to the unseen and unseeable—to long-lost species, to crucial ecosystems on the verge of collapse, and to icons of human civilization threatened by climate change. Each in their own way, these examples open our eyes, and our imaginations, to absences for which some human beings are responsible and about which we all might learn to care.

Memorializing *Ectopistes migratorius*

As we learned in chapter 2, names bear a unique rhetorical burden. To name someone or something is to announce, always in advance and in anticipation, a loss to come; the name is a promissory note of sorts, one which vows to safeguard its referent from the loss of both matter and memory.[23] Let us begin, then, with *Ectopistes migratorius*, the Latin name for a species of avifauna once dominant across much of central and eastern North America. On its website devoted to the species, the Smithsonian Institution notes that the "scientific name carries the connotation of a bird that not only migrates in the spring and fall, but one that also moves about from season to season to select the most favorable environment for nesting and feeding." Etymologically, "*Ectopistes* means 'moving about or wandering,' while *migratorius* means 'migrating.'"[24] Always on the move, this doubly itinerant species is better known as the passenger pigeon, a long-lost cousin of the ubiquitous rock pigeons (*Columba livia*) and mourning doves (*Zenaida macroura*) with whom a great share of the human population coexists today.

Unlike its extant kin, the passenger pigeon is extinct—and has been since the first day of September in 1914. On that date, in an aviary at the Cincinnati Zoo, the last surviving passenger pigeon, "Martha," perished.[25] Named after the first First Lady, Martha Washington, Martha the passenger pigeon's death was and was not remarkable. On the one hand, it is rare that we know the exact date that a species becomes extinct. Thus, the novelty of the circumstances is perhaps enough to warrant attention, and it is surely the case that Martha's story arouses sentiments that the anonymous deaths of so many other birds does not. On the other hand, however, billions of passenger pigeons preceded Martha in death, some of natural and some of unnatural causes. Focusing too much on the singular loss of an individual bird obscures the reality of extinction, which can never be reduced to a particular death. A single specimen does not, after all, a species make.[26] And in the case of the passenger pigeon in particular, the collective matters. "When Europeans first arrived in North America," the natural historian Joel Greenberg notes, "passenger pigeons likely numbered anywhere from three to five billion," making it "the most abundant bird on the continent, if not the planet."[27] Less than a century before Martha's death, the famed painter and naturalist John James Audubon could write that "the multitudes of Wild

Pigeons in our woods are astonishing." He was "struck with amazement" when, in 1813, he witnessed a flock of passenger pigeons flying over the Ohio River Valley; so immense was the avian assemblage that several days passed before it finally cleared the valley.[28] Within a century, such sights as Audubon witnessed would be gone forever, the space between infinity and finitude having collapsed under the weight of direct and indirect attacks on the passenger pigeon's ability to survive in its native range.

In the same account where he marvels at the sheer abundance of passenger pigeons, Audubon remarks that even though "they are killed in immense numbers . . . no apparent diminution occurs."[29] Writing prior to the publication of Charles Darwin's *On the Origins of Species*, Audubon struggled to fathom that *Ectopistes migratorius* could become extinct, particularly not as a result of human actions, however insidious and widespread they were. Although people indigenous to what is now called North America had peacefully coexisted with passenger pigeons for millennia, it took European settlers just a couple of centuries to push the species over the edge of extinction. Greeting the birds with a mixture of "wonder and hunger," the newly arrived settlers set about killing passenger pigeons with their bare hands, with poisons, with arrows, with bullets; they beat them with clubs and trapped them in nets.[30] They were eaten in immense quantities; their feathers were used to make pillows and other soft goods; they were shot for fun. A sketch by Smith Bennett appearing in the July 3, 1875, edition of *Illustrated Sporting and Dramatic News* well illustrates the strained relation among humans and passenger pigeons in the nineteenth century: in the top third of the image, a dense flock of pigeons darkens the sky; below, men with guns aimed skyward send bullets into the mass of birds. Here and there passenger pigeons fall from the sky, yet the flock hardly seems disturbed.[31] "Nothing," Audubon asserts in his 1831 *Ornithological Biography*, "but the gradual diminution of our forests [could] accomplish their decrease."[32] In the end, the newly arrived settlers diminished the forests, too. Oak, beech, chestnut, and other sylvan species upon whom the passenger pigeons depended for both food and habitat were felled by human hands or succumbed to infectious diseases, thus decreasing the species' capacity for survival.

Thirty-three years after Martha died, a gathering of avian enthusiasts—some who identified as ornithologists and others who adopted the humbler "birder" designation—met atop Sentinel Ridge overlooking the confluence of the Wisconsin and Mississippi rivers in southwestern Wisconsin's Wyalusing State Park. Organized by the Wisconsin Society for Ornithology, the assembly had traveled to the state park outside the small town of Prairie du Chien to witness the dedication of a monument to the passenger pigeon. Writing in the society's newsletter, aptly (and still) entitled *The Passenger Pigeon*, one Gertrude M. Scott recounts the events of May 11, 1947. After lunch, she writes,

> Members of the Society and many park visitors who were attracted by the unusual idea of erecting a monument to an extinct species of fauna, began gathering on Sentinel ridge, overlooking the broad Mississippi with its extensive wooded bottomlands, which is here swelled by the waters of the Wisconsin. The neat and imposing stone monument was the focal point, its face covered by a square of black cloth.[33]

Aldo Leopold, who we have met in previous chapters, was among the event's attendees. In the months leading up to the dedication ceremony, Leopold penned an essay for a pamphlet the society would disseminate during the festivities.[34] "This monument," Leopold laments, "symbolizes our sorrow. We grieve because no living man will see again the onrushing phalanx of victorious birds, sweeping a path for spring across the March skies, chasing the defeated winter from all the woods and prairies of Wisconsin."[35]

Rather serendipitously, I found myself on Sentinel Ridge seventy-two years to the day after the monument to the passenger pigeon was dedicated.[36] Having wound our way up the twisting, tree-lined lane, Brandon parked our car, and we walked the short distance from the gravel parking lot to the memorial, passing a series of Indian mounds along the way.[37] As we ambled toward the monument, we paused to read a sign positioned along the trail before we reached the memorial. "As you walk to the Passenger Pigeon Monument," the sign encourages passersby, "imagine huge flocks of pigeons flying up this river valley. They twist and turn as one. Together, they land to feed on acorns and other tree nuts. Together, they nest in colonies that stretch for miles." Perched on the edge of Sentinel Ridge, the memorial sat beneath mature oak trees, which cast shade and dappled light onto the memorial as American robins sang in the branches overhead. Flat stones had been placed on the ground, forming a floor. On three sides of the memorial, shallow stone walls lent a sense of enclosure and provided a place to sit, something against which to lean. From these walls rose a stout stone column, the monument's centerpiece, tapered faintly from base to apex. Set into the column at eye level was a bronze plaque, into which had been cast the impression of a single passenger pigeon alighting on an oak branch. Set in profile, the pigeon's slender form and piercing eyes were striking. "Dedicated to the last Wisconsin passenger pigeon shot at Babcock, Sept. 1899," the inscription read. "This species became extinct through the avarice and thoughtlessness of man."

When the memorial to the passenger pigeon was erected in 1947, people who could remember seeing the birds still lived. Though populations of passenger pigeons had already been severely reduced by the end of the nineteenth century, elders who witnessed the birds in their youth still existed. And so, to a certain extent, the monument may have kindled a living memory in its earliest days.[38] Today, however, visitors to the memorial possess no such memories of their own. The memorial's work consists, then, not of jogging visitors' personal memories but, rather, of constructing what some scholars call a "public memory," a remembrance of the past that is constructed to serve the needs of the present. As Carole Blair, Greg Dickinson, and Brian Ott note, "Public memories are constructed of rhetorical resources."[39] Put differently, public memories are invented, conjured, and imagined; they are fabricated with words, images, signs, stones, bronze, and much else besides. What we "know" about a place's history, for instance, is generated by the myriad of artifacts, stories, and symbols that conspire to "reveal" its past to us. What we colloquially call the "past" is thus an achievement, a complex rhetorical accomplishment, one that is far from neutral and always open to contestation.[40] Just as monuments in city squares draw our attention to some aspects of a place's history rather than others, so too does the monument to the passenger pigeon on Sentinel Ridge invite us to imagine the site in some ways rather than others. "If we listen and look for it," writes Jennifer Ladino, "the matter at memorials tells us how to feel, and it is insistent in its calls for our ethical attention."[41] Memorials make things matter.

Almost by definition, memorials bear some relationship to loss.[42] Often built in the wake of wars (e.g., the Vietnam Veterans Memorial) or traumatic events (e.g., the National September 11 Memorial and Museum), memorials summon attention to who or what has been lost. As such, memorials tend to be born of grief, and they are often meant to stimulate mourning. Indeed, as Marilyn DeLaure notes, "Memorials are arguably the most concrete, enduring, and (usually) officially sanctioned sites for public mourning."[43] This is not to say that everyone who visits a memorial will be moved to tears. Nor is it to say that invoking the dead or—in the case of the Passenger Pigeon Monument—the extinct always sets our grief into motion. Nevertheless, it is fair to say that the presence of memorials in particular places goes a long way toward generating the conditions of possibility for grief, including ecological grief. Memorials render invisible losses visible; they put before our eyes what we otherwise could or would not see, the hidden and imperceptible past, and in this way, memorials open us up to an encounter with absence itself. Since we cannot mourn what we do not know we have lost, such disclosures are significant.

The informational sign next to the Passenger Pigeon Monument instructs passersby to "imagine huge flocks of pigeons flying up this river valley." Below this injunction sits a full-color reproduction of Smith Bennett's drawing of men volleying bullets into a flock of passenger pigeons that darkens the sky above, thus providing visitors with a visual sense of what such "huge flocks" may have looked like. On the right side of the sign, the phrase "From Billions to None" is juxtaposed with a beautiful painting of three passenger pigeons alighting on oak branches, a visual echo of the monument's own impression. Although not every visitor will pause to read the interpretive sign or study the images, many will. While neither the text nor the images on the sign determine how visitors experience the memorial, they nevertheless influence the shape and feel of the encounter. If heeded, the provocation to imagine what is absent, to conjure an image of the disappeared birds as one looks out at the valley below and beyond, has the power to radically transform visitors' experience and understanding of place. To envision *what was* modifies and unsettles our sense of *what is*, for it disrupts the illusion that the present is a smooth emanation of the past, that what one sees today has always been so, that the abundance of life one witnesses is not, as it were, all there is or could be.

So it is on Sentinel Ridge. Standing at the memorial, one's attention is drawn away from the stone and bronze structure toward the horizon, toward the passenger pigeon's two great loves, "the fat of the land and the oxygen of the air."[44] The presence of the memorial imbues the landscape with a sense of sadness; looking out over the river valley below, one is struck both by the beauty of what is and the sorrow of what might have been. As Helen Whale and Franklin Ginn note, "absence and grief arise through perception of landscape, but also from the *knowledge* that it was once a place" where passenger pigeons lived, fed, mated, gave birth.[45] By insisting on the prior presence of the pigeons, by conjuring the birds in words, images, stone, and bronze, the sign and memorial invite visitors to reassess the abundance in front of and around them. The lush forests and wildflowers, the birds and mammals, the twining rivers below take on a different feel in the wake of this reassessment; they come to seem marked, or rather haunted, by an increasingly conspicuous loss. Indeed, the memorial encourages contemporary visitors to imagine the landscape as it once was, to see beyond (historically speaking) the present moment, to appreciate the dynamism of the land community through which they are passing. Simply put, the memorial demands that visitors acknowledge the land's historicity—the fact that it has a history, one that has been shaped as much by human as by inhuman forces. By imagining the past, one in which this valley is populated by, as the sign puts it, "limitless" passenger pigeons, visitors are called to view the world we have inherited, the earth we inhabit, as diminished—as less rich, less biologically diverse, less lively, and less interesting than it might otherwise have been. We come to see the horizon before us, ample though it is, as somehow wanting.

What the Passenger Pigeon Monument makes visible, then, is the precariousness of particular species and of land communities. By restoring to the landscape a sense of time and change, the memorial resists the impulse to take whatever is before us for granted and as an inevitability. If it evokes a sense of nostalgia in the process, this is surely not its only affective register, for memorials provoke multiple and complex feelings.[46] It also generates a sense of grievability, a sense that *what was* warrants our grief, our concern, and our care. The memorial attests to and conjures such concern, suggesting by its sheer existence that some humans have (and should continue to), as Leopold put it with this memorial in mind, "love what *was*."[47] And herein lies the rhetorical potency of this memorial: it persists in its place across time. Solid, it remains. Erect, it endures. "Perched like a duckhawk on this cliff," Leopold wrote in 1947, the memorial "will scan this wide valley, watching through the days and years."[48] Wind and rain and snow may, hundreds or thousands of years hence, erode it, but for now—and for the foreseeable future—it remains with us still, setting our imagination as well as our grief into motion.[49]

Walking back to the car, we again encounter the low-lying mounds whose presence an informational sign renders conspicuous. A schematic overhead drawing of the mounds atop Sentinel Ridge reveals a collection of twenty-eight distinct mounds—a mix of conical and linear mounds, as well as one bear-shaped effigy mound. Throughout Wisconsin, such mounds have dotted the hills and valleys for millennia. Some, though not all, of the mounds contain buried human remains; others may have possessed spiritual significance or marked a place as sacred. According to archaeologists Robert Birmingham and Amy Rosebrough, between fifteen and twenty thousand mounds existed when European and American settlers arrived in present-day Wisconsin.[50] Constructed by people of the Woodland Complex or Effigy Mound Culture, about whom we know little, the mounds atop Sentinel

Ridge are among the roughly four thousand or so surviving mounds in the state. "Time and erosion have obscured many mounds," writes geographer John Cross, "and modern agricultural plowing and urban development have obliterated" many others.[51] Today, extant mounds on private and public lands alike are protected by state and federal laws.

Whereas the Passenger Pigeon Monument invites visitors to imagine the billions of passenger pigeons that once moved through the lush forests and river valley one witnesses from Sentinel Ridge, the nearby mounds and attendant signage call attention to the peoples who once dwelled on this bluff and the riverbanks below. And just as the Passenger Pigeon Monument acknowledges the anthropogenic causes of the bird's extinction, one of the signs near the mounds obliquely implicates humans in the loss of many of the mounds: "Farming and development," the sign notes matter-of-factly, "have destroyed" all but a few thousand mounds in the state. This destruction is part of the history of Wyalusing State Park, which, the sign explains, "had 107 Indian Mounds in 1848"—the year that Wisconsin became the thirtieth state in the union. "Today the park has 64." Even as the sign calls attention to the destruction of mounds, though, it does not connect those losses to the broader processes of settler colonialism, land dispossession, or removal of Indigenous inhabitants that paved the way for "farming" and "development."[52] The sign avoids this history by narrowly focusing on the peoples who built the mounds and not mentioning the Ho-Chunk or other Indigenous peoples who have inhabited the area for hundreds of years, some of whom claim that their ancestors constructed the effigy mounds.[53] The sign avoids, in other words, the complicated history of settlement in which Native peoples were wrenched from their ancestral dwelling places, collective land was transformed into private property, and settlers were emboldened to irreparably alter natural and cultural landscapes.[54]

But it is not only the history of settler colonialism's catastrophic impacts on Native peoples that the signage atop Sentinel Ridge refuses to acknowledge. Despite the proximity of the Passenger Pigeon Monument and the Indian mounds, the signs draw no lines between the former and the latter.

They forge no connections between the ideological and material practices of domination that have historically underwritten both the attempted genocide of various Indigenous peoples and the extermination of more-than-human beings and ways of being. Surely the "avarice and thoughtlessness" responsible for the passenger pigeon's loss also played a part in the "destruction" of the mounds and in the devastation of Indigenous communities and ways of dwelling on the land. Such losses are not the same, but they are entangled. Atop Sentinel Ridge, the bear effigy mound to my left and the Passenger Pigeon Monument to my right, I cannot help but wonder what it would mean to imagine not only "huge flocks of pigeons flying up this river valley" but the Native peoples with whom they shared this place for hundreds, perhaps thousands, of years before European and American settlement. While the park does not explicitly encourage this kind of entangled imagining, what Casey Schmitt calls "reflexive layering," it could.[55]

Watching Melting

Let us move from rural Wisconsin to urban Copenhagen (and thence to Paris and London), from the extinction of a single species to the transformation of the cryosphere and, with it, the earth's climate. Days before the Intergovernmental Panel on Climate Change was set to present its "Fifth Assessment Report" in October 2014, twelve "enormous blocks of ice weighing 100 tonnes" were deposited on the square outside of Copenhagen's City Hall.[56] Roughly the size of a sedan, each of the twelve hunks of ice had been harvested from Nuup Kangerlua fjord near Nuuk, Greenland, and transported, first by boat and then by truck, to the Danish capital, where they were arranged into a circular formation. So assembled, the humongous chunks of sea ice assumed the look of a gigantic watch face, a time piece for a period of earth's history characterized by climate chaos and global warming.[57]

Conceived and produced by artist Olafur Eliasson and geologist Minik Rosing, the Copenhagen installation was aptly named *Ice Watch, 2014*. It was the first time Eliasson moved ice for art, but not the last. A year later, to coincide with the COP21 conference in Paris, *Ice Watch* reappeared on the Place du Panthéon. And, more recently, *Ice Watch* debuted at two sites in London—Bloomberg's European headquarters and the Tate Modern museum—in December 2018.[58] In an artists' booklet released at the Paris showing, readers learn that the "twelve immense blocks of ice are free-floating icebergs that have melted off from the Greenland ice sheet." Together, the booklet notes, the icebergs "present a tangible and immediate testimony of the dramatic effects of climate change."[59] While each iceberg melts over the course of days or weeks, it makes visible—"tangible and immediate"—what all too often occurs out of sight, out of mind. *Ice Watch* transports an ostensibly distant impact of climate change into the beating hearts of urban public life: the city square, the art museum, the corporate headquarters.

As many scholars have noticed, one of the challenges of mobilizing around climate change is the false, misleading notion that the effects of climate change are both temporally and spatially distant.[60] The dissonance continues. The Yale Program on Climate Change Communication found in 2019 that, while 67 percent of Americans "think global warming will harm future generations," only 48 percent "think global warming is already harming people in the U.S. now or within 10 years." Even fewer—only 42 percent—said they "think global warming will harm them personally."[61] Although a majority of people believe global warming is happening, they tend to assume that its impacts will be experienced somewhere else, at a later date, by people other than themselves. We have all heard, many of us for our entire lives, that as temperatures warm, ice will melt, oceans will rise, and our lives—especially, though not exclusively, the approximately three billion of us who live on or near a coastline—will be disrupted.[62] But the pace of those changes, rapid though they may be in geological terms, remains difficult for many to detect. As the Yale study illustrates, awareness of an ongoing crisis does not always foment feelings of urgency or immediate concern. When privileged people imagine that changes related to climate change, and the losses that follow, are distant in space and time, they can also continue to imagine that they will not be affected. Whatever else it is, climate change is a crisis of the imagination.

Just as we all need to become capable of imagining past and future losses, we also need to cultivate our collective capacity for imagining the very real losses occurring right now, in the ongoing present. This is especially true of physically remote losses, losses that seem to be taking place neither here nor now, but which nonetheless affect us. Most of us do not live near the Amazon rain forest, for instance, yet it is vital that we begin to imagine the immensity of what is being lost there as one of the earth's most diverse ecosystems is slashed and burned to make way for cattle farms. Similarly, though few of us live near the Greenland ice sheet, we all desperately need to be able to visualize the scale and the speed of the melting that threatens the ice with erasure. What happens there also happens here (wherever "here" happens to be). And yet, making the connection between *there and then* and *here and now* often demands a leap of the imagination and an ability to recognize links that are not always obvious. "The nagging *presence*," Helen Whale and Franklin Ginn note, "of that which threatens to disappear—a disappearance both happening and perpetually threatening—captures the imagination."[63]

Enter *Ice Watch*. Quite literally, the installation moves a seemingly distant crisis into the spheres of everyday, cosmopolitan life. After removing the icebergs from the fjord, Eliasson and Rosing's team transported the chunks of ice to their, as it were, final resting place: the place where they would finish melting. Ironically, the very global transportation network that contributes to global warming also makes it possible to relocate the massive icebergs. Having been severed from one context and sutured into another, the fragments of the Greenland ice sheet bridge the chasm between *here* and *there*, *then* and *now*, turning an otherwise easily neglected event into a forcefully present, proximate happening. Unlike the melting of the Greenland ice sheet, which too easily recedes into the background because it is happening "elsewhere," the melting icebergs are difficult to ignore precisely because they seem "out of place." The placement of the icebergs in Copenhagen, Paris, and London transgresses the usual boundaries that make it possible to disassociate "here" from "there," materially and symbolically enacting the first principle of ecology: all things are, indeed, connected. The fate of one place is the fate of all places.

It is not insignificant that the ice of *Ice Watch* was once part of the Greenland ice sheet. Eliasson and Rosing could have chosen to manufacture the ice themselves. Surely it would have been simpler logistically and cheaper financially to freeze water closer to the installation sites. The fact that the icebergs were connected to the Greenland ice sheet subtly but surely infuses their existence with what Walter Benjamin, in "The Work of Art in the Age of Its Technological Reproducibility," calls "aura." For Benjamin, aura is tied to realness and is embodied in the spectator's relation to the presence of something real. "To follow with the eye—while resting on a summer afternoon—a mountain range on the horizon or a branch that casts its shadow on the beholder," Benjamin explains, "is to breathe the aura of those mountains, of that branch."[64] One is, in a sense, touched by the presence of what one sees. To encounter *Ice Watch* is not to encounter an *image* of Greenland's ice sheet. Nor is it to encounter a *simulation* of glacial and sea ice melting. Instead, to encounter *Ice Watch* is to encounter the melting ice itself, for what one sees *is* the ice sheet, albeit in a fragmented form. If each iceberg functions synecdochically, a part standing in for the whole, they do so literally. Having been detached from the ice sheet by unnatural forces—first anthropogenic climate change, then Eliasson's prerogative—the icebergs nevertheless arrive intact, each one a genuine part of something larger, older, threatened, and in need of our concern and care.

It is for this reason, perhaps, that the creators of *Ice Watch* emphasize the origins and transportation of the icebergs. In the artists' booklet, a comment at the bottom of one page briefly tells the tale of how the hunks of ice ended up at the Place du Panthéon in Paris:

> Transport [of the ice was] organised by Group Greenland/Greenland Glacier Ice. The ice was collected by divers and dockworkers from the Royal Arctic Line and then transported in six refrigerated containers from Nuuk to

> Aalborg, Denmark, by container ship and to Paris by truck. Calculations for the project's carbon footprint have been carried out by Julie's Bicycle.[65]

And in both the artists' booklet and the website devoted to the project, numerous photos of the icebergs being collected and moved accompany images of the finished installation. In a series of photographs, we see someone pulling an iceberg behind their boat, a man in a hard hat loading an iceberg into a steel container, and another man operating a massive lift, upon which one of the translucent chunks of ice is precariously perched.[66] By documenting and displaying what might be called the "chain of custody" of the icebergs, Eliasson and Rosing repeatedly remind spectators that what they confront in *Ice Watch* is authentic, not a representation but a direct emanation of the source itself.

Ice Watch invites those who encounter it to imagine the melting of the Greenland ice sheet and, by extension, the melting of glaciers and sea ice across the globe, by exposing them to the materiality of such losses. Here, *Ice Watch* suggests, is what global warming looks like, feels like, smells like, sounds like, and even tastes like. As the ice melts, droplets of liquid water trickle down, one by one, forming small streams of water that dampen the shoes of passersby and form puddles on the pavement. Melting is not abstract; it is tangible, real, messy. You can, *Ice Watch* demonstrates, watch the melting in real time. It is not necessarily invisible, but it does need to be brought before our eyes. You can also hear the loss, as many commentators noted. "The icebergs are made up of compressed snow rather than frozen water, meaning that each block contains tiny little bubbles of air," one reporter explains. "Put your ear up against them," he suggests, "and you will hear a magical popping and fizzing sound."[67] Photographs from the installation show people doing just this. Indeed, the icebergs encourage interaction. People not only observe the melting visually, they also reach out and touch the icebergs, running their hands (and sometimes their tongues) across their slick surfaces. *Ice Watch* makes visible the materiality of melting. The installation underlines the reality of global warming, exposing its effects in ways that are palpable and accessible with our ordinary senses. It refuses the idea that climatic changes are out of sight and out of mind, forcing spectators to grapple with the undeniable fact of melting.

Precisely because *Ice Watch* invites and even encourages interaction, it also goes a long way toward exposing the real, yet often indiscernible, ways that we who live fossil-fueled lives are diminishing ecological and earth systems in seemingly remote places. Put differently, it helps us to imagine how our resource-intensive ways of life have downstream effects that we hardly notice because those impacts are felt most acutely beyond the realm of our everyday lives. In photographs and videos taken at the three *Ice Watch* exhibitions, one notices the extent to which passersby seem to be drawn in by the melting ice. And they do more than look; many also touch the icebergs. In one image, a dark-haired woman who is clad in all black lays her hands on one of the icebergs as she puckers up and kisses it. In another photograph, several young women place their fingertips on an iceberg while others snap photos with their smartphones. In yet another image, a man hoists a child up into an opening in an iceberg. Another photo shows two blond-haired boys licking the ice. Indeed, in nearly every image of people encountering *Ice Watch*, many, if not all, of the spectators are *touching* the ice in one way or another.

If we usually affect glacial melt indirectly through the burning of fossil fuels, *Ice Watch* stages a situation in which passersby can—and are enticed to—do so more directly. Indeed, if there is a metaphor at the heart of this public art display, it is that our desire to "touch," to handle, to grasp, to manipulate the ice is akin to our desire, or our proclivity, to manipulate our surroundings more generally. As photographs from the installation make obvious, "we" cannot keep our hands off the ice. As each person reaches out to hug, kiss, lay on, or otherwise touch the icebergs, they contribute—in a minute but undeniable way—to the deterioration of the ice in front of them and to the diminishment of the Greenland ice sheet. Indeed, touching *Ice Watch* makes dramatically visible what typically happens invisibly: human actions set the melting into motion in the first place and human actions will, in part, help to determine whether the melting will continue. Touching the ice, and watching others do so, summons attention to humans' responsibility for both directly and indirectly playing a role in the transformation and degradation of the natural world. For its part, *Ice Watch* connects specific human actions ("touching" writ small and large) to their consequences (melting and global warming), articulating humanity's—and, in the case of this installation, especially Europeans'—ways of being in the world to tangible ecological losses that are occurring here and there, in one's midst but also across the globe, in the present.

Visualizing connections between human actions and ecological consequences, *Ice Watch* bolsters our capacity for imagining not only ongoing losses but also, and crucially, our responsibility for them, if not for precipitating such losses, then certainly for how we respond to them. At the same time, passersby who touch the icebergs both form an intimate connection with the Greenland ice sheet (breathing in the trapped air it releases, for instance, and ingesting water molecules that have been frozen for tens of thousands of years) and actively, if not intentionally, imperil its existence. And, of course, engaging with *Ice Watch* is an individual and a collective experience. To this end, *Ice Watch* stages in miniature one of the troubling dimensions of life in the Anthropocene: we are increasingly aware not only of ecological and earthly losses but also of our varying degrees of culpability in perpetuating a culture that makes those losses inevitable. Drawing our attention to an ongoing loss, *Ice Watch* suggests that we might yet be moved by our grief for the melting ice to reimagine and remake our relations to the fossil-fueled ways of life threatening the Greenland ice sheet and other ice formations across the earth. We may not be able to stop these particular icebergs from melting, the installation asserts, but we may be able to prevent the worst from happening.

Picturing the End of the World

So far in this chapter we have visited a memorial that invites visitors to imagine an extinct species of birds that once dominated the central and eastern United States. We considered, too, a series of public art installations which make visible and tangible the ongoing loss of glacial ice, as well as the direct and indirect roles *Homo sapiens* plays in the diminishment of seemingly distant earth systems. If I am right that our capacity to grieve ecological losses depends, in part at least, on the composition of a visual field within which ecological losses appear *as losses*; and if I am correct in claiming that making earthly diminishment visible bolsters our ability to imagine and, thus, respond to losses past, present, and to come; then these examples suggest just some of the ways in which visual rhetoric can set our ecological grief into motion. In the next few pages, I train my sight on a final example, a project which makes visible the lose-ability not of another species or land community or precarious earth system, but, rather, of the all-too-human worlds modern humans have inherited.

As we dwell into the Anthropocene, Roy Scranton contends, "The greatest challenge we face is a philosophical one: understanding that this civilization is already dead."[68] That is, the most pressing test we now face is not, as is often assumed, scientific or technological in nature. We know that climate change is happening and that it is caused and exacerbated by the collective actions of billions of human beings who enjoy a fossil-fueled, resource-intensive way of life. And while scientists will continue to collect data and generate reports demonstrating this uncomfortable truth, the basics have already been well established.[69] Technology will inevitably factor into mitigation and adaptation efforts, but the problem is not primarily a technological one. Though some argue that the same human inventiveness and ingenuity that got us into the present predicament can also get us out of it, others—and here I include myself—believe that something other than new tools is in order.[70] Scranton's point is that what we need most is to confront, intellectually and emotionally, the fact that Western "civilization" as we have long known it will have to wither away and give rise to something different if humans hope to survive the Anthropocene. The infrastructures and ideologies that have underwritten "progress" in "developed" nations must cede ground to more ecocentric ideas and practices, including those carried on by Indigenous communities around the globe.[71] Thus, the greatest challenge of our time is coming to grips with the fact that peaceful earthly coexistence will require that we *lose* and, thus, that we *grieve* Western civilization as we know it.

Let us travel, finally, to the city of London—an icon of Western civilization if ever there was one. In *Postcards from the Future*, a series of images produced by artists Robert Graves and Didier Madoc-Jones (who call themselves "GMJ"), viewers are afforded a look at London as it might appear in the decades to come.[72] With photographs of iconic London structures and landscapes as their foundation, Graves and Madoc-Jones created composite images through collage, transposition, and other modes of digital editing and artistry. Thus does *Postcards from the Future* combine the familiar with the strange as mundane scenes of London are converted into dystopic and utopian visions of a future altered by climate change. In perhaps the most famous of the *Postcards*, we observe London from above as the sun is setting. In the foreground are Westminster Palace and Big Ben. Further away yet still prominent, the London Eye—the Ferris wheel that gives tourists a bird's eye view of London—summons attention to the middle third of the image as the city sprawls out in the distance as far as the eye can see. This recognizable scene is made strange, though, by the presence of water: the Thames River, which meanders through the center of both London and this image, has overflowed its barriers and inundated the city. In a more hopeful image from the collection, the Mall—a broad road connecting Buckingham Palace at one end and Trafalgar Square at the other—has been reimagined as a scene not just of royal power but of wind power, too. Where flag poles once flanked either side of the Mall, the artists have superimposed towering wind turbines, each one capped with the royal crest and decorated with a Union Jack flag. As these two examples demonstrate, the images comprising Graves and Madoc-Jones's *Postcards* project gesture toward diverse futures, some more hopeful than others. Whether the image depicts a catastrophic or promising time to come, however, each of the *Postcards* invites viewers to imagine an utterly transformed London. As the project's tagline, "Wish you were here?," makes plain, the question these *Postcards* pose is: Is this the sort of future you wish to inhabit?

Postcards from the Future was initially exhibited at the Museum of London from October 2010 to March 2011. Originally presented "as large scale back-lit transparencies," the oversized images were later displayed at the National Theatre in London from March to June 2011.[73] In addition to appearing in museums and galleries, the images also circulated in the popular and environmental media, appearing in the *Times*, the *Guardian*, *BBC News*, and *Grist*. During the Museum of London exhibition, the artists issued a call for other artists to create images of their own, and the competition that ensued attracted several hundred submissions, generating more attention for Graves and Madoc-Jones's original images.[74] Meanwhile, the *Postcards* persist in two other important forms: a website that features the images and brief descriptions thereof as well as a book of physical postcards available for purchase on the artists' website. Printed on cardstock and ready to be mailed, the miniature versions of the images bring the *Postcards* metaphor to its inevitable end. Like the postcards tourists send to friends and family members, Graves and Madoc-Jones's images invite recipients to imagine another place. Unlike those familiar epistolary media, though, the *Postcards* also make visible a time to come—a time whose precise characteristics are, in some sense, still to be determined.

The *Postcards* make visible futures that are at once fantastical and frightening. In one image, for instance, Piccadilly Circus is flooded with several feet of water. The usually bustling junction is submerged. Wind turbines rise out of the flooded Circus while dozens of solar panels grace the roofs of the surrounding buildings. Meanwhile, water lilies float on the placid water. The Shaftesbury Memorial Fountain at the center of Piccadilly Circus is nearly drowned; only the top of the Anteros statue breaks the water's surface. Notably, not a single person is visible in the scene. In another of the images, "Notting Hill Carnival," humans fill the frame. Onlookers young and old watch, smartphones and cameras in hand, as a parade of entertainers waving flags marches down the street. This otherwise familiar tableau is made strange by the fact that the spectators and performers alike are blue: every exposed bit of human skin has been superimposed with azure "sunscreen." Reminiscent in appearance of the Na'vi characters in James Cameron's 2009 film *Avatar*, the subjects in "Notting Hill Carnival" seem unearthly, out of place. As the artists explain on the project website, "A chance discovery of a shot of a young woman applying sun cream (shown at the bottom left of the image) catalysed the idea of transforming and uniting London's carnival audience in blue sunblock."[75] In yet another of their *Postcard* views, Graves and Madoc-Jones have transformed Parliament Square into sodden rice paddies. Haze obscures Westminster Palace and Big Ben in the background; in the foreground, faceless laborers carry out the back-breaking work of harvesting rice where London's power brokers once walked among statues of Winston Churchill, Abraham Lincoln, Nelson Mandela, Mahatma Gandhi, and others.

When Graves and Madoc-Jones originally exhibited their futuristic images in 2010 at the Museum of London, they were met with excitement and curiosity as well as hostility. Writing in the *Guardian*, for instance, George Marshall lambasted the work: "Not any one of these images reflects any real climate scenario for London," Marshall remarked. "They are pure science fiction." Marshall—who, along with Richard Sexton, founded the Climate Outreach Information Network in 2004, in an effort to transform climate change rhetoric—was not only frustrated by what he saw as a lack of fidelity to credible scientific predictions. He, like many other viewers, was disturbed by Graves and Madoc-Jones's visual treatment of climate refugees. "The greatest concern with this show," Marshall lamented, "is not that it parts with reality, but that it speaks all too well to real prejudices against immigrants."[76] One of the images in question portrays Buckingham Palace from above. Surrounded on all but one side by shanty homes, the sovereign's extravagant residence sharply contrasts with the nearly indistinguishable—because so closely congregated—homes of the newly arrived refugees. Marshall's concern, which I share, is that the image risks stoking fear rather than compassion by depicting a flood of immigrants "taking over" London's hubs of wealth and power. Another of the images, "Trafalgar Square Shanty," transports viewers to the depths of a shanty town assembled around Nelson's Column. While the iconic column rises up in the middle of the image, it now competes for attention with rustic structures on either side of a confined walking path where dark-skinned residents and customers interface. "Amid this 'new establishment,'" Graves and Madoc-Jones explain in their caption for the image, "Nelson's Column assumes a secondary, almost inconsequential significance."[77] In these and other images, Marshall detects hints of xenophobia and racism: portraying an influx of climate refugees does more to stoke anxieties about cultural transformations, he contends, than it does to enable viewers to empathetically welcome immigrants into Britain's political and cultural core.

While sympathetic to Marshall's criticisms, I want to emphasize another dimension of *Postcards from the Future*—not to downplay the project's shortcomings but, rather, to underline how the images gesture toward a different kind of ecological loss. What we see in Graves and Madoc-Jones's *Postcards* is the loss of the *status quo ante*, of the world as we inherited it. Faithful or not to scientific predictions, the *Postcards* make visible, each in their own way, the volatility and unpredictability of the future. Each image invites viewers to imagine a time to come that, although it bears a resemblance to the past, looks and feels remarkably different from the present. The view is at times alarming. The aerial image of an overflowing Thames may well strike fear in Londoners; inspired by photographs of New Orleans in the wake of Hurricane Katrina, "London as Venice" suggests chaos and disorder as the terrestrial world gives way to water. Blurred and broken boundaries are unsettling; when we can no longer tell where the river ends and the city begins, neither can we so neatly distinguish "nature" from "culture," the "human" from the "inhuman." Such boundaries, fiercely though they are protected, wither away in the Anthropocene. At other times, the view is less alarming than encouraging. In "Thames Tidal Power," for instance, we are again looking down upon an altered world; in this image, however, we see dozens of tidal turbines dotting the rippled River Thames. Just as other images in the series signal sustainable energy futures, so too does "Thames Tidal Power" picture a time to come when coal, oil, and natural gas have given way—either by choice or by necessity—to solar, wind, and hydro power. And while it is possible, perhaps even worthwhile, to point out when and where the images lapse into hyperbole, what matters most, I believe, is the way that these *Postcards* disturb viewers' abilities to take the continuity of the past, present, and future for granted.

Without wishing to draw a fine line between content and form, and without wishing to suggest that the content of Graves and Madoc-Jones's images is irrelevant, I nevertheless want to suggest that the *Postcards* matter less for *what* they represent than for *how* they invite viewers to imagine the end of the world as we know it. The scenarios that find expression in these images almost surely reflect the anxieties and aspirations of their makers, but the representational content of the *Postcards* is only one part of the story. It makes more sense, I believe, to understand them as "gestures of concern," rhetorical deeds that, as Chris Ingraham explains, "share . . . an expressive concern that acts as both a means and as an end because their most instrumental effects are exhausted in their expressivity."[78] Gestures of concern convey concern itself. When we send a "Be Well" card, for instance, we do not believe the card will heal our friend or family member; rather, the card expresses our interestedness in their condition and relays our concern for their health and well-being. Along these lines, the *Postcards* express the artists' interestedness in the future—and, in particular, with the pliability of the time to come. Indeed, the creative spirit which birthed these images attests to the responsibility we each bear to look, unflinchingly, into the future and to envision other ways of being in the world and on the earth with others. Graves and Madoc-Jones's images, however hyperbolic, insensitive, or misguided some of them may be, engender our responsibility by refusing to see our shared future as an inevitability.

In other words, by making visible possible futures, the postcards *make the future itself matter*. They make the time to come a matter of shared concern, something for which we are collectively responsible, not just for envisioning but for manifesting.[79] As Graves and Madoc-Jones's images make clear, whatever forms—and it will, of course, be plural—the future takes, loss is unavoidable. Not only will other species become extinct. Not only will land communities continue to change. Not only will ecosystems be further diminished. Not only will earth systems be disrupted. Civilization itself—whatever that means in each geographic, cultural, and political context—will also come undone. The world that we have inherited is, in the end, not a world that can be sustained, at least not in whole. As Scranton notes, "Human civilization has thrived in what has been the most stable climate interval in 650,000 years. Thanks to carbon-fueled industrial civilization, that interval is over."[80] In a sense, we who live fossil-fueled lives have lost the future that we were promised: a future characterized by perpetual growth, an unceasing linear "progress," and apparently limitless material abundance. As we lose one world, questions arise: What sorts of worlds will we imagine and construct? What from the past will we grieve in the Freudian sense of moving on, of letting go? What from the past will we cling to? What relations will we sever? And which will we refuse to relinquish?

Throughout this chapter, I have deliberately traveled to diverse geographical and rhetorical contexts. Rather than embracing eclecticism for eclecticism's sake, however, my aim has been to cast some light upon a multiplicity of rhetorical practices which, each in their own way, bring ecological losses past, present, and future before our eyes. The varied rhetorical practices I have considered in the preceding pages—timber signs, bronze memorials, public art, speculative images—operate on audiences in different ways, to be sure. What ties them together, though, is their capacity to conjure up what we have already lost (ancient eastern hemlocks and passenger pigeons), what we are presently losing (glacial ice and the stability of our earth systems), and what we still stand to lose in the decades and centuries to come (the post-industrial world we have inherited). What ties these various rhetorical practices together, in other words, is how they invite us to imagine loss as a defining characteristic of earthly coexistence—not as something to be willed away but, rather, as something with which we must grapple. Against the propensity to notice only what is, to neglect what has come before us, to ignore what we are destroying, and to act as though tomorrow will look the same as today, the sign along the walking path, the memorial on the ridge, the ice in the street, and the images on the postcards make visible precariousness itself. Take nothing for granted, they urge. Imagine how things once were, how they could have been, how they might yet become otherwise. Care about what has been lost, they insist, and care for what we are losing.

EPILOGUE

Caring Rhetorics

Meditations

On Saturday, August 17, 2019, nearly a hundred people clad in colorful parkas and wool caps trekked through dirt and scrambled over volcanic rocks as they climbed to the site of what once was Okjökull—a glacier (*jökull* in Icelandic) atop the Ok volcano northeast of Iceland's capital, Reykjavik. The group gathered to lament the loss of Okjökull, the first of Iceland's hundreds of glaciers to be declared "dead" due to anthropogenic climate change.[1] Satellite imagery taken in 1986 and 2019 reveals the transformation; where the glacier once sprawled, now only fragmented patches of compacted snow remain.[2] Now that the glacier has withered away, the site is known simply as Ok.[3]

The gathering atop Ok culminated in the unveiling of a bronze plaque, a marker that Cymene Howe and Dominic Bower—anthropologists at Rice University in Houston, Texas, both of whom research environmental issues and Icelandic culture—first imagined years earlier.[4] Affixed to a large boulder where the glacier once sprawled, the plaque itself features a "Letter to the Future," presented in both Icelandic and English:

Ok is the first Icelandic glacier to lose its status as a glacier.
In the next 200 years all our glaciers are expected to follow the same path.
This monument is to acknowledge that we know
what is happening and what needs to be done.
Only you know if we did it.

Ágúst 2019
415ppm CO2

Penned by Andri Snaer Magnason, a well-known Icelandic author, the brief missive on the memorial not only commemorates the loss of Okjökull; it also announces the fact that we understand both the causes of glacial melting and the responsibility we bear as a species to prevent further losses in Iceland and elsewhere. Written with present and future readers in mind, the plaque suggests that current generations can, and should, be judged on the degree to which we stave off such losses. In a couple hundred years, the plaque insists, our heirs will assess us based on our handling of the climate crisis. Hopefully, it implies, they will not find our response inadequate. Perhaps, the plaque insinuates, we will have been moved by the death and the memorialization of Okjökull—and perhaps other glaciers—to better care for the earth, our only home.

The English verb *to mourn* comes from the Gothic *maurnan*, "to worry" or "to concern oneself"; the Old High German *mornēn*, "to sorrow"; and the Old Saxon words *mornon* and *mornian*, "to be sorrowful or troubled." Its roots dig still deeper, however, reaching down to the Indo-European foundation of the word *memory*, where it connects with the Greek word *μέριμνα*, meaning "care" and "sorrow." Some now obsolete meanings of *to mourn* are instructive. It has meant "to be anxious or worried; to care; to have regard for." It has meant "to care about" and "to long to do something." And it has meant "to long (for)" and "to hanker *after*."[5]

To worry. To regard. To long for. To hanker. To care. In these and other deeds, we carry out the work of mourning, the work our grief sets into motion.

Always already entangled, mourning and caring mutually instigate, inform, and inflect one another. We grieve for who and what we care about. Our regard for, our love of, another moves us to grieve their loss; in their absence, we find ourselves bereaved. And yet, at the same time, we care for who and what we would mourn. When we recognize that the loss of the other would leave us bereft, deprived of a sustaining relation, we do whatever we can to prevent that loss from coming to pass. Or, at least, we become more capable of caring for others when we see them as worthy of our grief.

Our grief reflects to us our interconnection and interdependence, the fact that we are who and what we are only in relation to others. Our mourning shows us who and what matter, who and what mattered, to us. When we share our grief with others, it becomes a revelation: funerals, obituaries, memorials, and remembrances unconceal the ties that bind us, one to the other, the material and emotional links which tether us. But our acts of care—sometimes subtle, sometimes substantial—also reveal to us for whom and what we would mourn and, thus, who and what we will labor to safeguard against loss. A sense of anticipatory grief, a foreboding feeling of preemptory pain for a potential loss, animates our efforts to protect our fellow travelers and the land, ecosystems, and earth systems upon which we collectively depend.

On Saturday, April 27, 2019, protestors throughout Australia, Britain, Finland, France, New Zealand, Norway, the Netherlands, Sweden, and beyond laid their bodies down upon the ground to summon attention to the earth's sixth mass extinction event.[6] Strewn across shopping malls, art galleries, city streets, and museums, these protestors' prostrate bodies made visible and material the potential costs of maintaining the status quo. If we do not cultivate more ethical ways of dwelling on the earth with others, their vulnerable bodies announced, we run the risk of making the planet uninhabitable not only for our more-than-human kin but for our human heirs, too.[7]

Organized by Extinction Rebellion, an international movement that strives to "halt mass extinction and minimize the risk of social collapse," the April 27 die-ins in these countries were but one of the decentralized group's many actions aimed at both figuratively and literally disrupting the ceaseless flow of an unsustainable global fossil-fueled capitalist culture.[8] The pause these precarious bodies invite us to enact poses a series of questions: Would we mourn the loss a future in which humans find refuge on earth? Indeed, would we grieve the loss of the earthly conditions under which we have evolved with others and which we require if we are to survive and thrive? And will our anticipatory grief motivate us to alter the way we live today so that there might be a tomorrow for our own species and others?

Care is complicated. It is multiple, many-sided. Both a disposition and an activity, care entangles us in relations with others that are at once intellectual, emotional, and material.[9] We care about, we care for. Others care about and care for us. Care flows forth from the condition of interdependency. We are sometimes concerned with or by one another, and we sometimes work to support and sustain one another. As orientation and deed, care joins concern for the other with a willingness to put oneself in their service. Or, at least, care promises that one might lead to the other—that our concern for others might catalyze and underwrite efforts to safeguard the other against injury or loss. And, perchance, that we, too, might be sustained by the care of the other.

The work of earthly coexistence entails—indeed, requires—care of many kinds. If we wish to dwell more peacefully upon the earth with others, human and more-than-human alike, we will collectively need to become more caring. We will need to learn to care about those animals and plants and land communities and ecosystems that find themselves under threat of erasure today, and we will need to learn to care for these earthly others who so enrich our own lives and ways of being. But we will also need to acknowledge the extent to which we are cared for by the earth itself, a lesson I learn again and again as I read the work of Indigenous thinkers.[10] Considering all the ways that we are taken care of by what exceeds the human has a way of humbling us and generating the sense of gratefulness that undergirds our own acts of care and concern.

I have been reading, thinking, discussing, and writing about ecological rhetoric for more than a decade, initially as a member and leader of local conservation organizations in my community and then as an academic. Steeped in these issues, I know how easy it can be to focus myopically on what people—especially those of us of European descent—have done and are doing wrong; on humanity's long history of violence and exploitation; and on the ways that our communicative practices all too often prompt and exacerbate ecological crises. I have witnessed this tendency both outside and inside the academy. For instance, I entered the discipline of rhetoric two years after the journal *Environmental Communication* was established and J. Robert Cox described the fledgling scholarly community as one of "nature's crisis disciplines." In his widely cited article on the discipline's emergence and future, Cox argues that environmental communication scholars are obliged to both identify "distortions, ineptitudes, and system pathologies [that] occur in our communication about the environment" and to use our expertise to help solve the material crises which follow.[11] Since there will always be rhetorical disasters to diagnose and cure, it is unlikely that we will abandon Cox's perspective anytime soon.

Beyond uncovering and resolving crises in human communication about ecological problems, though, my experience within and outside of the academy has taught me that we also need to study and cultivate rhetorical practices by which we both express and induce care. While they may not use the language of "care" to frame their research, many scholars of ecological rhetoric already engage in this work.[12] Care furnishes a crucial supplement to the discipline's focus on crises, one which reminds us of the creative, restorative, earth-affirming powers of rhetoric. As Phaedra C. Pezzullo notes, we ought to attend to rhetorical forces that are "inspirational, healing, spiritual, and transformative in generative ways."[13] We should, that is, attend to how people rhetorically express their care for, as well as move others to care about, more-than-human species, land communities, ecosystems, and even the earth itself.[14]

On April 1, 2016, Ladonna Brave Bull Allard and other citizens of the Standing Rock Sioux Tribe established Sacred Stone Camp near the confluence of the Missouri and Cannonball Rivers, just north of the Standing Rock Indian Reservation in the present-day state of North Dakota. Founded to foster Native resistance to the Dakota Access Pipeline—a 1,172-mile-long oil pipeline beginning in North Dakota and unfurling across South Dakota and Iowa before its terminus in Illinois—Sacred Stone Camp began with a few dozen people but then expanded into several additional camps populated by more than ten thousand self-described "Water Protectors." "The protestors," writes Indigenous historian and activist Nick Estes (Kul Wicasa), "called themselves Water Protectors because they weren't simply against a pipeline; they also stood for something greater: the continuation of life on a planet ravaged by capitalism."[15]

Indeed, resistance to the Dakota Access Pipeline constitutes one chapter in a history of resistance that spans centuries in the He Sapa (Lakota for the Black Hills) and along the Mni Sose (Lakota for the Missouri River)—resistance to the US military's (unsuccessful) attempt to exterminate buffalo in the late nineteenth century; resistance to the US Bureau of Reclamation's and Corps of Engineers' damming of the Missouri River and subsequent flooding of Indigenous homelands in the mid-twentieth century; and resistance to the US government's termination policy during the 1940s, 1950s, and 1960s.[16] Resistance, in other words, to forced losses of companion species, of land, and of sovereignty. This history of deprivations, alongside the Lakota belief that "water is life" (*Mni Wiconi*), animated the Water Protectors' protests, ceremonies, and acts of civil disobedience. Though not all participants in the anti–Dakota Access Pipeline movement were aware of either these past losses or their colonial origins, the Standing Rock movement cannot be understood without considering these losses and the grief that they have generated over the years and across multiple generations.

But the movement to stop the pipeline from crossing the Missouri and Mississippi rivers, as well as Lake Oahe (one of the reservoirs created by the Pick-Sloan plan), was not only animated by a sense of past losses but also catalyzed by a feeling that the pipeline put something vital at risk of erasure—that the pipeline portended future losses that would, should they come to pass, warrant our grief. As Estes explains, "When land and water are taken and destroyed, so too is the possibility of a livable future."[17] Safeguarding a "livable future," a future in which human and more-than-human relatives will be capable of meeting their mutual responsibilities to one another, took many forms at the Standing Rock camps: chanting, teaching, dancing, cooking, standing, singing, talking. Through these and other acts, undertaken collectively and in public, the Water Protectors recognized the inherent value of human and more-than-human relations; revealed the labor, rhetorical and otherwise, required to sustain such relations; and demonstrated that the destruction of relations destroys more than just interchangeable parts in the present. Through these acts, in other words, the Water Protectors refused some relations and cared for others. To destroy relations is also to diminish possibilities for *becoming-with* others in times to come. Rejecting such destruction, diminishment, and loss, the Water Protectors held their ground until the last of the protestors was forcibly removed on February 23, 2017.

For better and for worse, "rhetoric" names an unruly, uncontainable collection of practices. For millennia people have debated what should count as rhetoric and, in the end, a variety of answers have been proposed—none of which can claim sole authority over what is, after all, an ancient and ever-evolving art.[18] Often villainized, often revered, rhetoric presents us always with a mixed bag: at one moment, rhetoric is aligned with deception; at another moment, rhetoric is framed as the beating heart of democratic societies. Political speeches that leave us feeling more (or less) united with our neighbors are considered rhetorical; so, too, though, are the late-night television advertisements that entice us to blow our paycheck on yet another gadget we neither need nor want. A big house, rhetoric plays host to all sorts of communicative practices.

If my undergraduate students' ideas about rhetoric are any indication of common knowledge, people tend to see rhetoric as synonymous with "persuasion." This makes some sense given the enduring association between rhetoric and Aristotle, who famously defined rhetoric as "the faculty of observing in any given case the available means of persuasion."[19] Unfortunately, though, persuasion is typically understood as a means of changing another person's mind by convincing them, via creative yet rational speech or writing, that they ought to modify their beliefs, commitments, or actions. As a result, people often want to assess rhetoric in instrumental terms by asking whether the rhetor successfully altered their audience's position on a particular issue. Yet, surely convincing listeners to replace Value A with Value B is not the only—or even the most common—way in which rhetors wield influence in the world.[20] As numerous rhetorical theorists have pointed out, rhetoric works in an array of ways.

One way to view rhetoric is to see it as a collection of worldmaking practices—a material-symbolic toolkit for making things matter, for imbuing specific objects, events, ideas, people, places, processes, and phenomena with salience, meaning, or value.[21] This is more or less the approach I have taken in this book and elsewhere, for I believe that long before we can debate policy or deliberate what we ought to do, we must share a sense of what matters, of what warrants our attention, of what is worthy of our concern and our care. The rhetorical practices I have considered in this book make more-than-human beings and ways of being matter in ways both subtle and explicit. The timber sign along the Alan Seeger Trail, for instance, merely gestures to the absence of the ancient eastern hemlock; it simply calls our attention to what was. By contrast, the wall of Latin and common names comprising Akroyd & Harvey's *Seeing Red.. Overdrawn* more overtly alerts audiences to the vulnerability of endangered species; it suggests that, without human intervention, these species will wither away, both metaphorically and literally. Not only do these artifacts make earthly entities matter, but they do so in ways that enable us to see the beings and ways of being to which they refer as grievable. It is from this sense of grievability—this sense that we would mourn these entities if they were lost—that we might derive the energy we need to labor on behalf of our fellow travelers.

Since rhetoric makes things matter in particular ways, the collective force of allied rhetorical practices is the composition of a world, which we can understand as a meaningful context. Thus, the question of what sorts of worlds we wish to inhabit, what sorts of worlds we hope to bequeath to our heirs, is in part a question about rhetoric, about the ways that we wield the material-symbolic materials (words, images, rocks, metals, bodies, ones and zeroes, et cetera) at hand to imbue some aspects of our shared reality with significance and with value beyond instrumentalism. With these tools at our disposal, we have the power to fashion worlds in which our fellow humans *care about* other beings and ways of being who find themselves at risk. With these tools at our disposal, we possess a powerful means of *caring for* the earth that bore and sustains us.

On Saturday, November 22, 2014, a little more than one hundred years after the last of the passenger pigeons perished, artist Emily Laurens conjured a specter of the species in sand at Llangrannog beach in Wales. Working at low tide with garden rakes, Laurens and her friends traced dozens of passenger pigeon silhouettes into the sand, "those fragments of rocks that sea and time [have] crumbled to near dust."[22] From a cliff above the beach, a place where photographs of the sand art installation were taken, the simulacra in the sand resembled the shadows of birds passing overhead in flight, visual reminders of the massive flocks that once darkened the skies of North America.

As the tide rose, each of Laurens's carefully crafted impressions washed away with the sea. In contrast with the Passenger Pigeon Monument in Wisconsin, which was built to endure for hundreds or thousands of years, Laurens's pigeons were designed to dissolve as the tides wash them away. Whereas the memorial persists, the sand art vanishes before our eyes, making visible the disappearance of the species itself.[23] In a blog post describing her sand art installation, Laurens writes:

> Our culture is focused on personal love between humans, extends someway to pets, and is obsessed with romantic love, but what about the love of place? The love of nature? The love of a lost species? Those feelings are not nurtured, not valued. We need to find new ways of expressing that love and commemorating the passing of places, ecosystems, and species.[24]

Both a century and an ocean separate Llangrannog beach and the forests and fields in North America where *Ectopistes migratorius* once flocked and flew. And yet, as the tides carry the pigeon silhouettes away, the love of which Laurens writes itself becomes visible, a love which tethers us to what was and that invites us to compose a less violent future, a future in which our more-than-human kin can take flight not only in our memories, a future in which the shadows on the sand move of their own accord.[25]

Essential to the difficult yet necessary work of earthly coexistence, through rhetoric we can care for the earth by generating the conditions under which others might come to care about our shared home. In this book, I have toured an array of rhetorical practices, from naming to archiving to making visible. My hope is that the examples that populate these pages will do more than play the role of "evidence" in the story I have been telling. My hope is that they might also furnish readers with models of caring rhetorics, models that could catalyze and inspire rhetorical practice in times to come.

When it comes to the work of earthly coexistence, I believe we must each begin wherever we find ourselves—in the communities, workplaces, schools, churches, and so on where we already happen to be. Indeed, since we are in some significant sense always already engaged in the work of earthly coexistence, we need not look for special places or times to care for the earth.[26] Every place, every time, is ripe. As a professor of rhetoric, I am privileged to work every day with smart and energetic students who are eager both to study rhetoric and to exercise their own creative muscles. For several years now, when I teach courses on topics like Ecological Communication or, more recently, Rhetoric and Earthly Coexistence, I toe the line between focusing on crisis and care, between attending to those rhetorical practices that tend to estrange humans from the earth and those that draw us into a sense of connection and intimacy with what exceeds the human.

The longer I teach these courses, the clearer it becomes that most students need not be convinced that climate change and other planetary changes pose a genuinely existential crisis. The sky has been falling since before they were born, and most of them have been hearing about ecological destruction, mass extinction, earthly diminishment (air and water pollution, for example), and innumerable other horrors their entire lives.[27] They seem well acquainted with stress, fear, anxiety, frustration, anger, despair, and worry, the terms they usually use to describe their feelings about the planetary emergency. Only after I talk about ecological grief do many of my students begin to open up about their own experiences of loss, their own sense of bereavement, their own feelings of grief. We discuss Sigmund Freud's "Mourning and Melancholia," consider Thom van Dooren's research on mourning crows, and talk through Ashlee Cunsolo and Karen Landman's argument that ecological grief can have both ethical and political implications.[28] And, by the end of the semester, students in my upper division Ecological Communication seminar compose and present eulogies for other animals and plants they/we have already lost, for species on the edge of extinction, for damaged landscapes, for irreparably harmed ecosystems, and for disrupted earth systems.[29] Over the course of a few weeks, my students go from never having heard the term "ecological grief" before to taking part in the creation of the rhetorical conditions under which others might be moved to view the earth as worthy of our grief.

Without overstating the influence of a single assignment or course, I would like to suggest that something significant happens when students write and share their eulogies at the end of the term. That something significant is this: by composing speeches that express and induce a sense of loss for what exceeds the human, they unite emotional intelligence—an understanding of how feelings shape our thoughts and our actions—with rhetorical efficacy—an ability to get things done, to evoke emotions and guide judgment, with the material-symbolic elements they have at hand.[30] I believe this union is the beating heart of a liberal education, an education which bequeaths us with the knowledge and the skills we need to live freely and to pursue the good life.[31] The good life, as my students' ecological eulogies end up reminding us, is a life lived in the company of others, human and more-than-human alike, for a life without others would quite literally be no life at all. And so, while the liberal arts prepare us for a life of freedom, this life is not—and cannot be—a life *free from* relations, connections, or dependencies. Since grief is a symptom of love, of connection, of association, its articulation has a way of showing us the extent to which our lives are entangled with others. Acknowledging how consequentially entwined we all are, of course, sets the work of earthly coexistence into motion, for it is only when we recognize our interdependence that we can begin to take responsibility for our shared fate. Delivering eulogies will surely not solve all our or the earth's problems, but it is one of many rhetorical practices that deliver us over to a world in which we come to see the earth as worthy of our care.

On Sunday, October 8, 2017, groups of people gathered at Harvard Forest in Petersham, Massachusetts, to witness artist David Buckley Borden and ecologist Aaron M. Ellison's collaborative site-specific art project, *Hemlock Hospice*.[32] Since its founding in 1907, scientists at Harvard Forest have studied and learned from the eastern hemlocks spread across the site's more than 4,000 acres.[33] The introduction of the hemlock woolly adelgid—an aphid-like bug about the size of a poppyseed that saps nutrients from the hemlocks—into the eastern United States in the 1950s, though, has put this sylvan species at risk of extinction.[34]

Visitors to *Hemlock Hospice* took a walk along a two-mile out-and-back trail through the woods. As they meandered through the forested landscape, they encountered eighteen distinct installations designed to draw attention to the demise of the hemlocks. At stop number nine, for instance, participants paused beside the "Memorial Woodshed," a wooden structure erected around the base of a dying eastern hemlock. Literally framing the vulnerable specimen, the structure invites consideration of and reflection on what will have disappeared, on what will have slipped out of the frame. Stop number twelve consists of a black "armband" wrapped around the trunk of a different hemlock, a sign that the tree itself is grieving. "A black Harvard Forest tree band," the tour brochure explains, "signifies that the wearer is in mourning and wishes to publicly commemorate a deceased family member, friend, or comrade." Thus are living hemlocks said to be grieving those hemlocks that have already succumbed to the adelgid. At the next stop along the trail, visitors viewed the "Exchange Tree," an artificial fallen hemlock made of large wooden spindles spiraling around a central pillar. Painted bright yellow, its branches summon attention to other trees in the forest that have already collapsed. To the "Exchange Tree" visitors were encouraged to attach a hand-written "message for yourself, for the hemlock, for the HWA, for others, or for the future of the forest."[35]

After years of reading, thinking, and planning, I began writing this book in earnest in early May 2020. The spring semester had just come to an end. I was gearing up to teach a rather intensive, three-week seminar on Earthen Rhetorics for a half dozen graduating seniors. My partner and I were readying ourselves to move across the country; with our home under contract and a closing date set, each passing day brought us closer to what we knew from experience would be a hectic couple of weeks. Meanwhile, we had barely left our house in more than six weeks. Aside from outings to the local co-op for groceries and, yes, the liquor store, we were taking the COVID-19 pandemic seriously and sequestering ourselves from other people. No longer driving to campus, no longer meeting with colleagues, no longer going out for dinner, no longer traveling, no longer wiling my mornings away in the café down the road, I finally found the time to put pen to paper.

In the time it has taken me to write this book, hundreds of thousands of Americans have perished as a result of contracting COVID-19.[36] Globally, more than four million people have reportedly died of the virus.[37] Writing about ecological grief during this pandemic, I could not help but notice the dearth of efforts to mourn those we have lost, on the one hand, and the range of vernacular mourning practices, on the other. Journalists and intellectuals alike have remarked on President Trump's refusal to publicly mourn people who have died of COVID-19, for instance.[38] In spite of Trump's failure to express or induce a national sense of grief, though, people have found ways to mark the loss: the *New York Times*, for example, created "Those We've Lost," a digital compendium of thousands of obituaries "designed to put names and faces to the numbers."[39] Meanwhile, the grassroots group COVID Survivors for Change staged a service in early October 2020 to remember the more than 200,000 Americans who had died of the virus at the time. Held on the Ellipse in Washington, DC, the centerpiece of the memorial was

20,000 empty chairs—each one representing ten people who died—facing toward the White House.[40]

What might these and other responses to the COVID-19 pandemic, and the many lives it has taken, teach us about ecological grief? First, it is always difficult to cope with an immense, even incalculable loss. Wars, genocides, natural disasters, and pandemics overwhelm us; the enormity of the holes they leave in the world often eludes our ability to grapple, to grieve. Second, the difficulty of grieving is compounded when we lack a sense of familiarity with who or what has been, or might be, lost. When we do not know the lost other, we struggle to mourn for them. And, third, if we believe it is important to grieve—if we believe that something significant happens when we confront our sorrow for the diminishment that each death, each disappearance, causes—we must come to terms with the fact that our capacity to grieve is itself the product of certain kinds of rhetorical labor. Before we can even begin to grieve, before we can even begin to recognize a loss as a loss, the rhetorical conditions that make those feelings probable must be put into place. That we ought to grieve is, as I hope to have demonstrated in this book, no guarantee that we can or will. Our capacity to grieve is a hard-won accomplishment.

But the COVID-19 pandemic furnishes another unfortunate yet valuable lesson for those of us who are interested in ecological grief. When death, disappearance, destruction, diminishment, and disruption result from active violence or passive neglect, the entities that are responsible for precipitating such losses—whether governments or corporations—tend to behave as though nothing, in fact, has been lost. To do otherwise is to concede that one's actions have severed a significant tie, that one's conduct has diminished our shared world. And, furthermore, to do otherwise is to admit that someone or something significant has been lost and, thus, that the world has been irretrievably altered. Under conditions like these, the act of publicly grieving—of avowing that a loss has occurred—is a powerful form of political and ethical resistance, one which insists both that we care about who and what we have already lost and that we care for who and what we might yet lose.[41]

On Monday, January 23, 2012, in the mountains north of Salt Lake City, Utah, director Jeff Orlowski's Emmy Award–winning documentary film *Chasing Ice* debuted at the Sundance Film Festival.[42] *Chasing Ice* follows photographer James Balog and a team of researchers, technicians, and assistants to Greenland, Iceland, and Alaska, where—under the auspices of Balog's "Extreme Ice Survey"—they set up dozens of cameras to record the melting and calving of glaciers. After struggling to make the cameras work in harsh environments and inclement weather, Balog's team collected tens of thousands of still photographs that, when organized into a chronological sequence, reveal changes to the glaciers. Balog's time-lapse series exposes the severity, intensity, scale, and speed of glacial loss; as the photos advance, we bear witness to the steady "retreat" of ice that took millennia to accumulate.

Of course, *Chasing Ice* is about the glaciers—about what is happening to them and about the forces that set those alterations into motion. The film is also about Balog, who is portrayed in familiar terms: in Balog we have a good-looking, masculine, compassionate, adventure-seeking, heroic protagonist—a "good guy" who is committed to changing how we dwell on earth by teaching us about the effects our actions have on distant places. There is, though, a third narrative: *Chasing Ice* is, in the end, a documentary about documentation itself, a film which lays bare the labor involved in attending to, recording, gathering, and sharing shifts in earthly phenomena. And, in this sense, it is a movie about rhetoric, about the ways humans make things matter, about how we care *for* more-than-human others by moving—or attempting to move—our fellow travelers to care *about* what exceeds the human. As we watch *Chasing Ice*, we not only watch the melting glaciers; we also witness the work which bringing such faraway phenomena, such seemingly distant losses, into public view requires.

Odd though it may seem, writing about ecological grief has infused me with a sense of hope. Although we have good reasons to feel anxious, worried, angry, fearful, frustrated, and skeptical about the prospects of earthly coexistence, attending to some of the ways in which humans express their own ecological grief and induce it in others provides us with a necessary reminder that there are other ways of dwelling on earth. When we recall that grief and mourning are symptoms of love, of intimacy, and of relation, we begin to see them—and experience them—not only through the lenses of pain and sorrow and privation, as many of us have been taught, but as crucial indications of connection across difference.[43] A world in which we can and would mourn beyond the human, after all, is a world in which we (will) have loved capaciously.

Studying the rhetorical conditions of our capacity to grieve beyond the human also fills me with a feeling of hopefulness. Indeed, for me, studying rhetoric is an inherently hopeful undertaking for it affirms, again and again, that the world can always be otherwise. Though we inherit worlds not of our making, we are not powerless to reimagine and remake them. The study of rhetoric is, at least in part, the study of how people creatively compose the worlds they wish to share with others, human and more-than-human alike. We may not always—or even often—agree on what our shared worlds ought to look and feel like. What matters more than consensus, however, is that we understand we are capable of shaping, if not determining, the conditions under which we live.

Throughout this book, my critical efforts have been aimed at calling attention to some of the ways that people are working to fashion worlds more hospitable for earthly coexistence.[44] We have seen how naming other living beings, other species, features of the land, and geological epochs enables us to maintain relationships with what we have already lost and to anticipate what we might yet lose. We have also seen how observing, recording, and archiving earthly phenomena—avian migration and flowering, to name but two—shed light on historical losses and transformations as well as help us prepare to mourn losses to come. And we have seen how practices of making ecological losses visible—from erecting memorials to creating speculative images of climate change–altered cityscapes—catalyze our capacity to imagine and respond to losses, past, present, and future.

And so, let us learn—and learn to utter—the names of our more-than-human kin and kind. Let us recognize that each name gestures, always in advance, toward a unique loss to come. And let us learn to attend to what is happening around us. Let us commit even some small portion of our time to opening our eyes, our ears, and our whole bodies to the vibrant, teeming earth that bore and sustains us. Let us share what we see and hear with others; let us construct the archives we need to tell new and different stories about earthly coexistence. And let us exercise our capacity to make the invisible visible, to let what has been unseen be seen, to unconceal what we stand to lose. Let us look at what worlds we have wrought and envision what we might yet make together with others, human and more-than-human alike. Let us feel ourselves becoming more connected to other beings and ways of being, to land communities, to places, to ecosystems, to the earth itself. Let us compose and maintain worlds in which we see what exceeds the human as worthy of our grief, our concern, and our care.

Rhetorically and otherwise, let us take care.

Notes

Prologue. Loss: A Reckoning

1. This attitude toward loss is characteristic of what Potawatomi scholar Kyle Powys Whyte calls "epistemologies of crisis." See Kyle Powys Whyte, "Against Crisis Epistemology," in *Routledge Handbook of Critical Indigenous Studies*, ed. Brendan Hokowhitu et al. (London: Routledge, 2020), 53–57.
2. Quoted in Frank B. Linderman, *Plenty-Coups: Chief of the Crows* (Lincoln, NE: Bison, 2002), 169. For an incisive reading of Plenty-Coup's remark, see Jonathan Lear, *Radical Hope: Ethics in the Face of Cultural Destruction* (Cambridge, MA: Harvard University Press, 2006).
3. For accounts of settler colonial violence against Native peoples and lands in the present-day United States, as well as narratives of resilience and survivance in the wake of such losses, see Winona LaDuke, *All Our Relations: Native Struggles for Land and Life* (Chicago: Haymarket, 2015); Winona LaDuke, *Recovering the Sacred: The Power of Naming and Claiming* (Chicago: Haymarket, 2016).
4. Nick Estes, *Our History Is the Future: Standing Rock versus the Dakota Access Pipeline, and the Long Tradition of Indigenous Resistance* (London: Verso, 2019), 134; see also Michael L. Lawson, *Dammed Indians: The Pick-Sloan Act and the Missouri River Sioux, 1944–1980* (Norman: University of Oklahoma Press, 1982).
5. See R. Alfred Vick, "Cherokee Adaptation to the Landscape of the West and

Overcoming the Loss of Culturally Significant Plants," *American Indian Quarterly* 35, no. 3 (2011): 394–417.

6. Core Writing Team, Rajendra K. Pachauri, and Leo A. Meyer, "Climate Change 2014: Synthesis Report" (Geneva: Intergovernmental Panel on Climate Change, 2014), https://www.ipcc.ch/site/assets/uploads/2018/02/SYR_AR5_FINAL_full.pdf.
7. See Timothy Morton, *Being Ecological* (Cambridge, MA: MIT Press, 2018).
8. For something of a compendium of losses and transformations tied to climate change, see David Wallace-Wells, *The Uninhabitable Earth: Life after Warming* (New York: Tim Duggan, 2019).
9. The consequences of human action are always unpredictable, but the complexity of the ecological and earth systems into which we now routinely act renders human action especially fraught today. For a myriad of examples, see Elizabeth Kolbert, *Under a White Sky: The Nature of the Future* (New York: Crown, 2021).
10. Dina Gilio-Whitaker, *As Long as Grass Grows: The Indigenous Fight for Environmental Justice from Colonization to Standing Rock* (Boston: Beacon, 2019), 138.
11. Henry David Thoreau, *Walden; Or, Life in the Woods* (Boston: Ticknor and Fields, 1854); Aldo Leopold, *A Sand County Almanac and Sketches Here and There* (New York: Oxford University Press, 1949).
12. Native Americans' decisions to engage in activities such as fossil-fuel extraction need to be seen within the broader contexts of settler colonialism and of compromised conditions of coexistence. See Gilio-Whitaker, *As Long as Grass*, 69–72.
13. For two examples of Native American tribes that have faced difficult ecological decisions, see Winona LaDuke, "Three Affiliated Tribes at a Crossroads: Which Energy Path?," *Indian Country Today*, April 6, 2006, https://indiancountrytoday.com/archive/three-affiliated-tribes-at-a-crossroads-which-energy-path; Winona LaDuke, "Crow and Lummi, Dirty Coal and Clean Fishing," *Indian Country Today*, January 15, 2014, https://indiancountrytoday.com/archive/crow-lummi-dirty-coal-clean-fishing.
14. Kathryn Yusoff, "Aesthetics of Loss: Biodiversity, Banal Violence, and Biotic Subjects," *Transactions of the Institute of British Geographers* 37, no. 4 (2011): 579.
15. For an example situated in the Canadian context, see Jennifer Clary-Lemon, "Examining Material Rhetorics of Species at Risk: Infrastructural Mitigations as Non-Human Arguments," *Enculturation: A Journal of Rhetoric, Writing, and Culture*

32 (2020), http://enculturation.net/material_rhetorics_species_at_risk.

16. My previous work on these topics includes Joshua Trey Barnett, "Naming, Mourning, and the Work of Earthly Coexistence," *Environmental Communication* 13, no. 3 (2019): 287–99; Joshua Trey Barnett, "Grievable Water: Mourning the Animas River," in *Water, Rhetoric, and Social Justice: A Critical Confluence*, ed. Casey Schmitt, Christopher S. Thomas, and Theresa Castor (Lanham, MD: Lexington, 2020), 273–91; Joshua Trey Barnett, "Vigilant Mourning and the Future of Earthly Coexistence," in *Intimate Relations: Communication (in) the Anthropocene*, ed. Vail Fletcher and Alexa Dare (Lanham, MD: Lexington, 2021), 13–33.
17. Linda Hogan, *Dwellings: A Spiritual History of the Living World* (New York: W. W. Norton, 1995), 115.
18. Aldo Leopold, "The Round River," in *A Sand County Almanac with Essays on Conservation from Round River* (New York: Ballantine, 1966), 197.
19. In chapter 4, I dwell on Eliasson's *Ice Watch* installation. See *Ice Watch, 2014*, Olafur Eliasson, accessed August 19, 2020, https://olafureliasson.net/archive/artwork/WEK109190/ice-watch.
20. LaDuke recounts some of her work on the White Earth Reservation in present-day Minnesota in LaDuke, *All Our Relations*, 115–34.
21. I adopt the notion of an "image of thought" from Gilles Deleuze, who contends that all thinking implies an image of thought premised on ideas of what "everybody knows" and of what "no one can deny." See Gilles Deleuze, *Difference and Repetition*, trans. Paul Patton (New York: Columbia University Press, 1994), 129–67.
22. Thom van Dooren, *Flight Ways: Life and Loss at the Edge of Extinction* (New York: Columbia University Press, 2014), 18.
23. Robin Wall Kimmerer, *Braiding Sweetgrass: Indigenous Wisdom, Scientific Knowledge, and the Teachings of Plants* (Minneapolis: Milkweed, 2013), 190.
24. Leopold, *A Sand County Almanac*, 204.
25. Douglas Burton-Christie, "The Gift of Tears: Loss, Mourning, and the Work of Ecological Restoration," *Worldviews* 15, no. 1 (2011): 31.
26. For a discussion of kinship relationships built on these commitments and values, see Whyte, "Against Crisis Epistemology," 57–59.
27. During the past two decades, grievability has become an organizing concept in Butler's

social, ethical, and political theorizing. See, for instance, Judith Butler, *Antigone's Claim: Kinship between Life and Death* (New York: Columbia University Press, 2000); Judith Butler, *Precarious Life: The Powers of Mourning and Violence* (London: Verso, 2004); Judith Butler, *Frames of War: When Is Life Grievable?* (London: Verso, 2009); Judith Butler, *Notes Toward a Performative Theory of Assembly* (Cambridge, MA: Harvard University Press, 2015).

28. I adopt this phrasing from Ashlee Cunsolo and Karen Landman, "To Mourn beyond the Human," in *Mourning Nature: Hope at the Heart of Ecological Loss and Grief*, ed. Ashlee Cunsolo and Karen Landman (Montreal: McGill-Queen's University Press, 2017), 3–26.
29. Animal Clock, 2020, http://animalclock.org.
30. "Roadkill Statistics," *High Country News*, February 7, 2005, https://www.hcn.org/issues/291/15268.
31. Marcel Huijser et al., "Wildlife–Vehicle Collision Reduction Study: Report to Congress" (McLean, VA: US Federal Highway Administration, 2008), 11, https://www.fhwa.dot.gov/publications/research/safety/08034/index.cfm.
32. See, for instance, Jacques Derrida and Elizabeth Roudinesco, "Violence against Animals," in *For What Tomorrow . . . A Dialogue*, trans. Jeff Fort (Stanford, CA: Stanford University Press, 2004), 62–76; Donna Haraway, *When Species Meet* (Minneapolis: University of Minnesota Press, 2008), 285–301.
33. Beverly Peterson Stearns and Stephen C. Stearns, *Watching, from the Edge of Extinction* (New Haven, CT: Yale University Press, 1999), x.
34. Elizabeth Kolbert, *The Sixth Extinction: An Unnatural History* (New York: Henry Holt, 2014), 15.
35. van Dooren, *Flight Ways*, 12.
36. Luis W. Alvarez et al., "Extraterrestrial Cause for the Cretaceous-Tertiary Extinction," *Science* 208, no. 4448 (1980): 1095–1108.
37. Kolbert, *The Sixth Extinction*, 73–74.
38. Gerardo Ceballos et al., "Accelerated Modern Human-Induced Species Losses: Entering the Sixth Mass Extinction," *Science Advances* 1, no. 5 (2015): 4.
39. "Population, Total," World Bank, 2019, https://data.worldbank.org/indicator/SP.POP.TOTL.

40. "Global Greenhouse Gas Emissions Data," US Environmental Protection Agency, September 13, 2019, https://www.epa.gov/ghgemissions/global-greenhouse-gas-emissions-data.
41. "Background and History," International Union for Conservation of Nature Red List, 2020, https://www.iucnredlist.org/about/background-history.
42. See Mark S. McClure, Scott M. Salom, and Kathleen S. Shields, "Hemlock Woolly Adelgid" (Morgantown, WV: USDA Forest Health Technology Enterprise Team, December 1996); Annie Paradis et al., "Role of Winter Temperature and Climate Change on the Survival and Future Range Expansion of the Hemlock Woolly Adelgid (*Adelges tsugae*) in Eastern North America," *Mitigation and Adaptation Strategies for Global Change* 13, no. 5–6 (2008): 541–54.
43. Richard B. Primack, *Walden Warming: Climate Change Comes to Thoreau's Woods* (Chicago: University of Chicago Press, 2014), 29.
44. John J. Wiens, "Climate-Related Local Extinctions Are Already Widespread among Plant and Animal Species," *PLoS Biology* 14, no. 12 (2016): 9–12.
45. Wiens, "Climate-Related Local Extinctions," 2.
46. Primack, *Walden Warming*, 63–64.
47. Quoted in Primack, *Walden Warming*, 57.
48. Edward O. Wilson, *Half-Earth: Our Planet's Fight for Life* (New York: Liveright, 2016), 14.
49. "Deforestation and Forest Degradation," World Wildlife Fund, 2020, https://www.worldwildlife.org/threats/deforestation-and-forest-degradation.
50. "Forest Habitat," World Wildlife Fund, 2020, https://www.worldwildlife.org/habitats/forest-habitat.
51. For an anthropological perspective on the entwinement of forests and thinking, see Eduardo Kohn, *How Forests Think: Toward an Anthropology beyond the Human* (Berkeley: University of California Press, 2013).
52. On the topic of the imbrication of forests and human culture, see Robert Pogue Harrison, *Forests: The Shadow of Civilization* (Chicago: University of Chicago Press, 1993); Richard Fortey, *The Wood for the Trees: One Man's Long View of Nature* (New York: Alfred A. Knopf, 2016).
53. Thomas E. Dahl and Susan-Marie Stedman, "Status and Trends of Wetlands in the

Coastal Watersheds of the Conterminous United States, 2004 to 2009" (US Fish and Wildlife Service and National Marine Fisheries Service, 2013), 1, https://www.fws.gov/wetlands/documents/status-and-trends-of-wetlands-in-the-coastal-watersheds-of-the-conterminous-us-2004-to-2009.pdf.

54. Dahl and Stedman, "Status and Trends of Wetlands," 20–21.

55. Thomas E. Dahl, "Wetlands Losses in the United States, 1780's to 1980's" (Washington, DC: US Department of the Interior, Fish and Wildlife Service, 1990), 1, https://archive.epa.gov/water/archive/web/pdf/wetlands-losses-in-the-united-states-1780s-to-1980s.pdf.

56. I learned most of what I know about the relationship between soil health and industrial agriculture from the writings of Wes Jackson, who founded The Land Institute in Salina, Kansas, in 1976. See Wes Jackson, *New Roots for Agriculture* (Lincoln: University of Nebraska Press, 1985); *Altars of Unhewn Stone: Science and the Earth* (Wooster, OH: Wooster, 1987); *Becoming Native to This Place* (Berkeley, CA: Counterpoint Press, 1994); *Consulting the Genius of the Place: An Ecological Approach to a New Agriculture* (Berkeley, CA: Counterpoint Press, 2010).

57. "Large 'Dead Zone' Measured in Gulf of Mexico," US National Oceanic and Atmospheric Administration, August 1, 2019, https://www.noaa.gov/media-release/large-dead-zone-measured-in-gulf-of-mexico.

58. Carol Turley et al., "Hot, Sour, and Breathless: Oceans Under Stress" (Plymouth Marine Laboratory, UK Ocean Acidification Research Programme, European Project on Ocean Acidification, Mediterranean Sea Acidification in a Change Climate Project, Scripps Institution of Oceanography at UC San Diego, OCEANA, 2013), https://www.pml.ac.uk/pmlsite/media/PML-Media/Documents/Oceans-under-stress-Nov13.pdf.

59. Turley et al., "Hot, Sour, and Breathless," 3.

60. Jérémie Mouginot et al., "Forty-Six Years of Greenland Ice Sheet Mass Balance from 1972–2018," *Proceedings of the National Academy of Sciences* 116, no. 19 (2019): 9239.

61. Michon Scott and Kathryn Hansen, "Sea Ice," NASA Earth Observatory, September 16, 2016, https://earthobservatory.nasa.gov/features/SeaIce.

62. Ed Struzik, "How Thawing Permafrost Is Beginning to Transform the Arctic," *Yale Environment 360* (blog), January 21, 2020, https://e360.yale.edu/features/

how-melting-permafrost-is-beginning-to-transform-the-arctic.

63. LaDuke, *All Our Relations*, 148.
64. Leopold, *A Sand County Almanac*, 209–10.
65. Donna Haraway, *Staying with the Trouble: Making Kin in the Chthulucene* (Durham, NC: Duke University Press, 2016), 39.
66. Jeremy David Engels, *The Art of Gratitude* (Albany: State University of New York Press, 2018), 7.

Chapter 1. Ecological Grief: A Rhetorical Achievement

1. Glenn A. Albrecht, *Earth Emotions: New Words for a New World* (Ithaca, NY: Cornell University Press, 2019), 1.
2. See Martin Heidegger, *Being and Time*, trans. Joan Stambaugh (Albany: State University of New York Press, 2010), 130–43.
3. Within the interdisciplinary environmental humanities, numerous scholars have considered the relationship between feelings and earthly coexistence. See, for instance, Albrecht, *Earth Emotions*; Joshua Trey Barnett, "Irrational Hope, Phenological Writing, and the Prospects of Earthly Coexistence," *Communication and Critical/Cultural Studies* 16, no. 4 (2019): 382–91; Joshua Trey Barnett and David Charles Gore, "Dwelling in the Anthropocene: Notes from Lake Superior," *Ethics & the Environment* 25, no. 1 (2020): 19–49; Joshua Trey Barnett, "Ecological Moods: Feeling Our Way into Thinking Like Aldo Leopold," *ISLE* 28, no. 3 (2021): 1155–81; Kyle Bladow and Jennifer Ladino, eds., *Affective Ecocriticism: Emotion, Embodiment, Environment* (Lincoln: University of Nebraska Press, 2018); Heather Houser, *Ecosickness in Contemporary U.S. Fiction: Environment and Affect* (New York: Columbia University Press, 2014); Tim Jensen, *Ecologies of Guilt in Environmental Rhetorics* (London: Palgrave Macmillan, 2019); Dolly Jørgensen, *Recovering Lost Species in the Modern Age: Histories of Longing and Belonging* (Cambridge, MA: MIT Press, 2019); Jennifer K. Ladino, *Memorials Matter: Emotion, Environment, and Public Memory at American Historical Sites* (Reno: University of Nevada Press, 2019); Renee Lertzman, *Environmental Melancholia: Psychoanalytic Dimensions of*

Engagement (London: Routledge, 2015); Robin Wall Kimmerer, *Braiding Sweetgrass: Indigenous Wisdom, Scientific Knowledge, and the Teachings of Plants* (Minneapolis, MN: Milkweed, 2013); Timothy Morton, "The Dark Ecology of Elegy," in *The Oxford Handbook of the Elegy*, ed. Karen Weisman (Oxford: Oxford University Press, 2010), 251–71; Timothy Morton, *Being Ecological* (Cambridge, MA: MIT Press, 2018); Kari Marie Norgaard, *Living in Denial: Climate Change, Emotions, and Everyday Life* (Cambridge, MA: MIT Press, 2011); Rosemary Randall, "Loss and Climate Change: The Cost of Parallel Narratives," *Ecopsychology* 1, no. 3 (2009): 118–29; Sarah Jaquette Ray, *A Field Guide to Climate Anxiety: How to Keep Your Cool on a Warming Planet* (Oakland: University of California Press, 2020); Deborah Bird Rose, *Wild Dog Dreaming: Love and Extinction* (Charlottesville: University of Virginia Press, 2011); Thom van Dooren, *Flight Ways: Life and Loss at the Edge of Extinction* (New York: Columbia University Press, 2014); Edward O. Wilson, *Biophilia: The Human Bond with Other Species* (Cambridge: Harvard University Press, 1984).

4. For a powerful example of the fallacy that knowledge alone is sufficient to spark environmental care, see Norgaard, *Living in Denial*, 64–73.
5. Jensen, *Ecologies of Guilt*, 155.
6. This is a common idea within conservationist discourse. See, for instance, Leath Tonino and Michael Soulé, "We Only Protect What We Love: Michael Soulé on the Vanishing Wilderness," *The Sun Magazine*, April 2018, https://www.thesunmagazine.org/.
7. Aldo Leopold famously learned this lesson while living and working in the American Southwest. It was there that he learned to "think like a mountain," which for Leopold meant thinking like an ecosystem. After watching a wolf die, Leopold began to question his position on predator species, realizing that his love of "game" animals like deer prevented him from respecting the shared conditions of earthly life. See Aldo Leopold, "Thinking Like a Mountain," in *A Sand County Almanac and Sketches Here and There* (New York: Oxford University Press, 1949), 129–33.
8. See, for instance, Peter Coates, *Strangers on the Land: American Perceptions of Immigrant and Invasive Species* (Berkeley: University of California Press, 2007), 151–64.
9. In a study of captive whooping crane breeding programs, Thom van Dooren describes what he calls "violent care," the commingling of "intimate care" with "domination,

coercion, and abandonment of others." Van Dooren, *Flight Ways*, 92.

10. Renee Lertzman powerfully demonstrates this in her book on environmental melancholia. As she puts it, we cannot "presume that how we behave is a straightforward expression of our concerns, beliefs, and values." See Lertzman, *Environmental Melancholia*, 145.
11. "Emotional intelligence" names the capacity to understand, manage, and express one's emotions. See Daniel Goleman, *Emotional Intelligence: Why It Can Matter More Than IQ* (New York: Bantam, 1995).
12. For one characteristic take on this distinction, see "Grief vs. Mourning: What's the Difference?," The Recovery Village, January 4, 2021, https://www.therecoveryvillage.com/mental-health/grief/related/grief-vs-mourning/.
13. For example, see Ann Cvetkovich, *Depression: A Public Feeling* (Durham, NC: Duke University Press, 2012). On rhetorical approaches to public feelings, see Jamie Landau and Bethany Keeley-Jonker, "Conductor of Public Feelings: An Affective-Emotional Rhetorical Analysis of Obama's National Eulogy in Tucson," *Quarterly Journal of Speech* 104, no. 2 (2018): 166–88.
14. My argument resonates with Tim Jensen's claim that, "Social pedagogies explicitly and implicitly instruct who is and who is not considered worthy of grief, boundaries that are built and enforced through myriad mechanisms." See Jensen, *Ecologies of Guilt*, 129.
15. Kriss Kevorkian, "Environmental Grief: Hope and Healing" (PhD diss., Union Institute and University, 2004), 2.
16. Ashlee Cunsolo and Neville R. Ellis, "Ecological Grief as a Mental Health Response to Climate Change–Related Loss," *Nature Climate Change* 8 (2018): 275.
17. Jennifer Atkinson, "Addressing Climate Grief Makes You a Badass, Not a Snowflake," *High County News*, May 29, 2018, https://www.hcn.org/.
18. Fakiha Baig, "Being Sad for the Planet: Edmontonians Share Experiences with Ecological Grief," *CBC News*, July 23, 2019, https://www.cbc.ca/news/.
19. Dan McDougall, "'Ecological Grief': Greenland Residents Traumatised by Climate Emergency," *Guardian*, August 12, 2019, https://www.theguardian.com/.
20. Jordan Rosenfeld, "Facing Down 'Environmental Grief,'" *Scientific American*, July 21, 2016, https://www.scientificamerican.com/; Ellen Airhart, "Taking Action Can Cure Your Climate Grief," *Scientific American*, March 23, 2021, https://www.

scientificamerican.com/.

21. Eric Holthaus, "If You're Suffering from Climate Grief, You're Not Alone," *Grist* (blog), October 15, 2018, https://grist.org/article/climate-grief-un-ipcc-report/.
22. Paige Cockburn, "Climate Grief Expected to Be Widespread Soon but It's Still Not Openly Acknowledged," *ABC News*, August 18, 2020, https://www.abc.net.au/.
23. Michaela Cavanagh, "It's Time to Talk about Ecological Grief," *Undark*, January 10, 2019, https://undark.org/2019/01/10/its-time-to-talk-about-ecological-grief/.
24. Thomas Attig, *How We Grieve: Relearning the World*, 2nd ed. (Oxford: Oxford University Press, 2011), 106.
25. Grief transforms who we "are." As Judith Butler puts it, "Perhaps we can say that grief contains the possibility of apprehending a mode of dispossession that is fundamental to who I am." In other words, the very possibility of loss exposes us to the fact that we are constituted in and by our relations with others. We cannot, therefore, help but become otherwise in the wake of a loss. See Judith Butler, *Precarious Life: The Powers of Mourning and Violence* (London: Verso, 2004), 28.
26. See Darcy L. Harris, ed., *Non-Death Loss and Grief: Context and Clinical Implications* (New York: Routledge, 2020).
27. David Redmalm, "Pet Grief: When Is Non-Human Life Grievable?," *Sociological Review* 63, no. 1 (2015): 19–35; Margo DeMello, ed., *Mourning Animals: Rituals and Practices Surrounding Animal Death* (East Lansing: Michigan State University Press, 2016).
28. Joshua Trey Barnett, "Vigilant Mourning and the Future of Earthly Coexistence," in *Communicating in the Anthropocene: Intimate Relations*, ed. Alexa Dare and C. Vail Fletcher (Lanham, MD: Lexington, 2021), 13–33; Jørgensen, *Recovering Lost Species*; Van Dooren, *Flight Ways*.
29. Helen Whale and Franklin Ginn, "In the Absence of Sparrows," in *Mourning Nature: Hope at the Heart of Ecological Loss and Grief*, ed. Ashlee Cunsolo and Karen Landman (Montreal: McGill–Queen's University Press, 2017), 92–116.
30. Joshua Trey Barnett, "Grievable Water: Mourning the Animas River," in *Water, Rhetoric, and Social Justice: A Critical Confluence*, ed. Casey Schmitt, Christopher S. Thomas, and Theresa Castor (Lanham, MD: Lexington, 2020), 273–91.
31. Dahr Jamail, *The End of Ice: Bearing Witness and Finding Meaning in the Path of*

Climate Change (New York: New Press, 2019); Nadine Marshall et al., "Reef Grief: Investigating the Relationship between Place Meanings and Place Change on the Great Barrier Reef, Australia," *Sustainability Science* 14 (2019): 579–87.

32. Cunsolo and Ellis, "Ecological Grief"; Ashlee Cunsolo et al., "'You Can Never Replace the Caribou': Inuit Experiences of Ecological Grief from Caribou Declines," *American Imago* 77, no. 1 (2020): 31–59.
33. David Abram, *The Spell of the Sensuous: Perception and Language in a More-than-Human World* (New York, NY: Vintage, 1996), 22.
34. See Douglas Burton-Christie, "The Gift of Tears: Loss, Mourning, and the Work of Ecological Restoration," *Worldviews* 15, no. 1 (2011): 30.
35. Cunsolo and Ellis, "Ecological Grief," 275.
36. Winona LaDuke, *All Our Relations: Native Struggles for Land and Life* (Chicago: Haymarket, 2015), 148.
37. I have only touched the proverbial "tip of the iceberg" of traumas inflicted on Indigenous peoples in the present-day United States. See also Roxanne Dunbar-Ortiz, *An Indigenous Peoples' History of the United States* (Boston: Beacon, 2014).
38. For instance, see Katherine Meloche, "Mourning Landscapes and Homelands: Indigenous and Non-Indigenous Peoples' Ecological Griefs," *Canadian Mountain Network* (blog), August 30, 2018, https://canadianmountainnetwork.ca/mourning-landscapes-and-homelands-indigenous-and-non-indigenous-peoples-ecological-griefs/.
39. LaDuke, *All Our Relations*, 140.
40. LaDuke, *All Our Relations*, 141–43.
41. Nick Estes, *Our History Is the Future: Standing Rock versus the Dakota Access Pipeline, and the Long Tradition of Indigenous Resistance* (London: Verso, 2019), 98.
42. Jonathan Lear considers this historical example in detail, with a special emphasis on how the Crow responded to and ultimately survived the devastation of the buffalo. See Jonathan Lear, *Radical Hope: Ethics in the Face of Cultural Destruction* (Cambridge, MA: Harvard University Press, 2006).
43. For other examples, see LaDuke, *All Our Relations*; Winona LaDuke, *Recovering the Sacred: The Power of Naming and Claiming* (Chicago: Haymarket, 2016).
44. In addition to documenting Inuit responses to ecological loss in academic research, Cunsolo also directed a short film based on her work with Nunatsiavut communities.

See Ashlee Cunsolo Willox, *Attutauniujuk Nunami/Lament for the Land*, 2014, http://www.lamentfortheland.ca/film/.

45. Quoted in Cunsolo and Ellis, "Ecological Grief," 277.
46. Quoted in Cunsolo and Ellis, "Ecological Grief," 277.
47. See Aldo Leopold, *A Sand County Almanac and Sketches Here and There* (New York: Oxford University Press, 1949), 44–50, 95–101, 108–12, and 129–33.
48. Aldo Leopold, "The Round River," in *A Sand County Almanac with Essays on Conservation from Round River* (New York: Ballantine, 1966), 197.
49. Phyllis Windle, "The Ecology of Grief," *BioScience* 42, no. 5 (1992): 364.
50. David R. Foster, "Hemlock's Future in the Context of Its Past," in *Hemlock: A Forest Giant on the Edge*, ed. David R. Foster (New Haven, CT: Yale University Press, 2014), 9; for a rhetorical analysis of this collection of essays, see Joshua Trey Barnett, "From Ecological Grief to *Gelassenheit*: Rhetorics of Sentiment and Science in *Hemlock*," *Environmental Communication* 15, no. 6 (2021): 783–97.
51. Lesley Head and Theresa Harada, "Keeping the Heart a Long Way from the Brain: The Emotional Labour of Climate Scientists," *Emotion, Space and Society* 24 (2017): 34–41.
52. Kimmerer, *Braiding Sweetgrass*.
53. Timothy A. C. Gordon, Andrew N. Radford, and Stephen D. Simpson, "Grieving Environmental Scientists Need Support," *Science* 366, no. 6462 (2019): 193.
54. Cunsolo and Ellis, "Ecological Grief," 275; Margo DeMello, "Introduction," in *Mourning Animals: Rituals and Practices Surrounding Animal Death*, ed. Margo DeMello (Minneapolis: University of Minnesota Press, 2016), xxi; see also Kenneth J. Doka, ed., *Disenfranchised Grief: Recognizing Hidden Sorrow* (Lexington, MA: Lexington, 1989).
55. See Catriona Mortimer-Sandilands, "Melancholy Natures, Queer Ecologies," in *Queer Ecologies: Sex, Nature, Politics, Desire*, ed. Catriona Mortimer-Sandilands and Bruce Erickson (Bloomington: Indiana University Press, 2010), 338.
56. Stef Craps, "Introduction: Ecological Grief," *American Imago* 77, no. 1 (2020): 3.
57. Susan Clayton, Christie Manning, and Caroline Hodge, "Beyond Storms and Droughts: The Psychological Impacts of Climate Change" (Washington, DC: American Psychological Association and ecoAmerica, 2014), 25.
58. Clayton, Manning, and Hodge, "Beyond Storms and Droughts," 33.

59. See Susan Clayton Whitmore-Williams et al., "Mental Health and Our Changing Climate: Impacts, Implications, and Guidance" (Washington, DC: American Psychological Association, Climate for Health, and ecoAmerica, 2017), https://www.apa.org/news/press/releases/2017/03/mental-health-climate.pdf.
60. Atkinson, "Addressing Climate Grief."
61. Judy Fahys, "First Step to 'Eco-Grieving' over Climate Change? Admit There's a Problem," National Public Radio, April 22, 2017, https://www.npr.org/2017/04/22/524557600/first-step-to-eco-grieving-over-climate-change-admit-theres-a-problem.
62. Lisa Kretz, "Emotional Solidarity: Ecological Emotional Outlaws Mourning Environmental Loss and Empowering Positive Change," in *Mourning Nature: Hope at the Heart of Ecological Loss and Grief*, ed. Ashlee Cunsolo and Karen Landman (Montreal: McGill–Queen's University Press, 2017), 258–91.
63. We need further empirical research on both the prevalence of ecological grief and on its symptoms. Surveys of wide swaths of people could be undertaken to assess how many people experience ecological grief, how that grief manifests in their daily lives, whether and how they transform that grief into mourning practices, and what factors of their everyday lives may make them more or less likely to experience ecological grief.
64. Elizabeth Kübler-Ross and David Kessler, *On Grief and Grieving: Finding the Meaning of Grief through the Five Stages of Loss* (New York: Scribner, 2005), 227.
65. See, for example, Roland Barthes, *Mourning Diary*, trans. Richard Howard (New York: Hill and Wang, 2010), 102–4.
66. Readers of Judith Butler will recognize the structure of this formulation, for it is a paraphrase and reworking of Butler's remark that, "We are undone by each other. And if we're not, we're missing something." See Butler, *Precarious Life*, 23.
67. See Andrew Mark, "Don't Organize, Mourn: Environmental Loss and Musicking," *Ethics & the Environment* 21, no. 2 (2016): 51–77.
68. Douglas Crimp, "Mourning and Militancy," *October* 51 (1989): 5.
69. Ray, *A Field Guide*, 8.
70. Ray, *A Field Guide*, 126.
71. Van Dooren, *Flight Ways*, 140.
72. In their introduction to *Mourning Nature*, Ashlee Cunsolo and Karen Landman

develop this point in detail. See Ashlee Cunsolo and Karen Landman, "Introduction: To Mourn beyond the Human," in *Mourning Nature: Hope at the Heart of Ecological Loss and Grief*, ed. Ashlee Cunsolo and Karen Landman (Montreal: McGill–Queen's University Press, 2017), 11–15; also see Ashlee Cunsolo et al., "Ecological Grief and Anxiety: The Start of a Healthy Response to Climate Change?," *The Lancet* 4, no. 7 (2020): e261–63.

73. Sigmund Freud, "Mourning and Melancholia," in *Collected Papers: Volume IV*, ed. Joan Riviere (London: Hogarth Press, 1953), 153.
74. Freud, "Mourning and Melancholia," 154.
75. For a useful description and critique of Freud's liberatory thesis, see Tammy Clewell, "Mourning beyond Melancholia: Freud's Psychoanalysis of Loss," *Journal of the American Psychoanalytic Association* 52, no. 1 (2004): 43–67.
76. Freud, "Mourning and Melancholia," 159.
77. Freud, "Mourning and Melancholia," 160.
78. See Jacques Derrida, "By Force of Mourning," in *The Work of Mourning*, trans. Pascale-Anne Brault and Michael Naas (Chicago: University of Chicago Press, 2001); R. Clifton Spargo, *The Ethics of Mourning: Grief and Responsibility in Elegiac Literature* (Baltimore: Johns Hopkins University Press, 2004); Barnett, "Vigilant Mourning."
79. Sigmund Freud, *The Ego and the Id: The Standard Edition*, ed. James Strachey, trans. Joan Riviere (New York: W. W. Norton, 1960), 24.
80. Jacques Derrida develops this point across a number of texts. See, for instance, Jacques Derrida, *The Politics of Friendship*, trans. George Collins (London: Verso, 1997); Jacques Derrida, "The Taste of Tears," in *The Work of Mourning*, ed. Pascale-Anne Brault and Michael Naas (Chicago: University of Chicago Press, 2001), 107–10; Derrida, "By Force of Mourning"; also see Pascale-Anne Brault and Michael Naas, "To Reckon with the Dead: Jacques Derrida's Politics of Mourning," in *The Work of Mourning*, by Jacques Derrida (Chicago: University of Chicago Press, 2001), 1–30; Joan Kirkby, "'Remembrance of the Future': Derrida on Mourning," *Social Semiotics* 16, no. 3 (2006): 461–72; Judith Butler, *The Psychic Life of Power: Theories in Subjection* (Stanford, CA: Stanford University Press, 1997).
81. Butler, *Precarious Life*, 23.
82. Cunsolo and Landman, "Introduction," 11.

83. See Judith Butler, *Frames of War: When Is Life Grievable*? (London: Verso, 2009); Barnett, "Grievable Water."

84. See Judith Butler, *Notes toward a Performative Theory of Assembly* (Cambridge, MA: Harvard University Press, 2015).

85. See George Yancy and Judith Butler, "What's Wrong with 'All Lives Matter,'" *New York Times*, January 12, 2015, https://opinionator.blogs.nytimes.com/2015/01/12/whats-wrong-with-all-lives-matter/.

86. See Cunsolo and Landman, "Introduction," 22–24.

87. See Kübler-Ross and Kessler, *On Grief and Grieving*, 1–5.

88. Morton, "The Dark Ecology," 254; Jessica Marion Barr, "Auguries of Elegy: The Art and Ethics of Ecological Grieving," in *Mourning Nature: Hope at the Heart of Ecological Loss and Grief*, ed. Ashlee Cunsolo and Karen Landman (Montreal: McGill–Queen's University Press, 2017), 197; see also Patricia Rae, "Double Sorrow: Proleptic Elegy and the End of Arcadianism in 1930s Britain," *Twentieth Century Literature* 49, no. 2 (2003): 246–75; Sebastian F. Braun, "Mourning Ourselves and/as Our Relatives: Environment as Kinship," in *Mourning Nature: Hope at the Heart of Ecological Loss and Grief*, ed. Ashlee Cunsolo and Karen Landman (Montreal: McGill–Queen's University Press, 2017), 83–84.

89. It is telling that Carson inaugurates her book with a line she attributes to theologian Albert Schweitzer: "Man has lost the capacity to foresee and forestall. He will end by destroying the earth." See Rachel Carson, *Silent Spring* (Boston: Houghton Mifflin, 1962), iv.

90. See Butler, *Frames of War*, 13–15; James Stanescu, "Species Trouble: Judith Butler, Mourning, and the Precarious Lives of Animals," *Hypatia: A Journal of Feminist Philosophy* 27, no. 3 (2012): 567–82; Barnett, "Grievable Water."

91. See Butler, *Frames of War*, 15–32.

92. Cunsolo and Landman, "Introduction," 16.

93. Soren C. Larsen and Jay T. Johnson, *Being Together in Place: Indigenous Coexistence in a More Than Human World* (Minneapolis: University of Minnesota Press, 2017), 79.

94. Larsen and Johnson, *Being Together in Place*, 82.

95. See Larsen and Johnson, *Being Together in Place*, 83–85.

96. For a partial history, see "School History," Haskell Indian Nations University, 2017,

https://www.haskell.edu/about/history/.

97. Larsen and Johnson, *Being Together in Place*, 84–85.

98. Larsen and Johnson, *Being Together in Place*, 87.

99. David Charles Gore, *The Voice of the People: Political Rhetoric in the Book of Mormon* (Provo, UT: Brigham Young University, 2019), 200.

100. Leopold, *A Sand County Almanac*, 48.

101. For instance, see Leopold, *A Sand County Almanac*, 223–26.

102. Plato, *Gorgias*, trans. Walter Hamilton and Chris Emlyn-Jones, rev. ed. (New York: Penguin, 2004); Plato, *Phaedrus*, trans. Christopher Rowe (New York: Penguin Classics, 2005).

103. Plato, *Gorgias*, 29, 30, 30, 31.

104. See Aristotle, "Rhetoric," in *The Complete Works of Aristotle*, ed. Jonathan Barnes, trans. W. Rhys Roberts, vol. 2 (Princeton, NJ: Princeton University Press, 1984), 2152–2269.

105. Plato, *Phaedrus*, 44.

106. Aristotle, "Rhetoric," 2155.

107. In a long essay entitled "Plato's Pharmacy," Jacques Derrida reconsiders Socrates' comparison of rhetoric to a "pharmakon" in the *Phaedrus*. As Derrida explains, although the Greek word "pharmakon" is often translated as "drug" or "poison," it can also mean "cure." Thus does Derrida discover in Plato's *Phaedrus* the basis for a more ambivalent orientation toward rhetoric, one that acknowledges the risks and promises of oratory and writing. See Jacques Derrida, *Dissemination*, trans. Barbara Johnson (Chicago: University of Chicago Press, 1981), 61–172.

108. For an early artifact of the migration from literary studies to the study of public address, see Herbert A. Wichelns, "The Literary Criticism of Oratory," in *Studies in Rhetoric and Public Address in Honor of James Albert Winans*, ed. A. M. Drummond (New York: Century, 1925), 181–216.

109. John Louis Lucaites and Celeste Michelle Condit, "Introduction," in *Contemporary Rhetorical Theory: A Reader*, ed. John Louis Lucaites, Celeste Michelle Condit, and Sally Caudill (New York: Guilford, 1999), 7–8.

110. See Edwin Black, *Rhetorical Criticism: A Study in Method* (Madison: University of Wisconsin Press, 1965).

111. In 1970, the National Development Project on Rhetoric convened more than

forty leading rhetorical scholars to assess the state of rhetorical studies and to make suggestions about its future. Among their recommendations was the need to grapple with the twentieth century's new technological landscape, to widen the scope of what counts as rhetorical theory and practice, to reconsider what we mean by rationality, and to make invention central to rhetorical inquiry. See Lloyd F. Bitzer and Edwin Black, eds., *The Prospect of Rhetoric: Report of the National Development Project* (Englewood Cliff, NJ: Prentice-Hall, 1971).

112. Phaedra C. Pezzullo, "Unearthing the Marvelous: Environmental Imprints in Rhetorical Criticism," *Review of Communication* 16, no. 1 (2016): 25–42.

113. For example, see Jenny Edbauer, "Unframing Models of Public Distribution: From Rhetorical Situation to Rhetorical Ecologies," *Rhetoric Society Quarterly* 35, no. 4 (2005): 5–24; Michael Calvin McGee, "Text, Context, and the Fragmentation of Contemporary Culture," *Western Journal of Communication* 54, no. 3 (1990): 274–89.

114. See Book II, sections 1–11, of Aristotle, "Rhetoric," 2194–2213.

115. For recent examples of scholarship focused on the connections between rhetoric and feeling, see Jeremy David Engels, *The Politics of Resentment: A Genealogy* (University Park: Pennsylvania State University Press, 2015); Jeremy David Engels, *The Art of Gratitude* (Albany: State University of New York Press, 2018); Lisa M. Corrigan, *Black Feelings: Race and Affect in the Long Sixties* (Jackson: University Press of Mississippi, 2020); Chris Ingraham, *Gestures of Concern* (Durham, NC: Duke University Press, 2020).

116. My definition of rhetoric resonates with Thomas Farrell's: "Rhetoric is the art, the fine and useful art, of making things matter." See Thomas B. Farrell, "The Weight of Rhetoric: Studies in Cultural Delirium," *Philosophy & Rhetoric* 41, no. 4 (2008): 470.

117. On the notion of "world," see chapter 3 of Heidegger's *Being and Time*. There, he makes clear that our perceptions and understandings of "nature" emerge from within worlds. See Heidegger, *Being and Time*, 63–110.

118. John Durham Peters, *The Marvelous Clouds: Toward a Philosophy of Elemental Media* (Chicago: University of Chicago Press, 2015), 51.

119. For discussions of rhetoric as a means of constituting subject positions, see Maurice Charland, "Constitutive Rhetoric: The Case of the *Peuple Québécois*," *Quarterly Journal of Speech* 73, no. 2 (1987): 133–50; Raymie McKerrow, "Critical Rhetoric: Theory

and Praxis," *Communication Monographs* 56, no. 2 (1989): 91–111; Barbara Biesecker, "Michel Foucault and the Question of Rhetoric," *Philosophy & Rhetoric* 25, no. 4 (1992): 351–64; Kevin DeLuca, "Articulation Theory: A Discursive Grounding for Rhetorical Practice," *Philosophy & Rhetoric* 32, no. 4 (1999): 334–48.

120. Engels, *The Art of Gratitude*, 11.

121. Ashlee Cunsolo and Neville Ellis use the word "natural" to characterize ecological grief in their seminal article. See Cunsolo and Ellis, "Ecological Grief," 275.

122. Just as rhetoric composes the conditions under which mourning becomes possible, so too can it constrict the time and space of mourning. See, for instance, Barbara Biesecker, "No Time for Mourning: The Rhetorical Production of the Melancholic Citizen-Subject in the War on Terror," *Philosophy & Rhetoric* 40, no. 1 (2007): 147–69.

123. Jacques Derrida, *The Animal That Therefore I Am* (New York: Fordham University Press, 2008), 20.

124. Michele Kennerly, "Getting Carried Away: How Rhetorical Transport Gets Judgment Going," *Rhetoric Society Quarterly* 40, no. 3 (2010): 269.

125. In making this argument, I follow Phaedra Pezzullo's lead in characterizing ecological rhetoric as a "care discipline." See, for instance, Phaedra C. Pezzullo, "Environment," in *Oxford Research Encyclopedia of Communication*, ed. Jon F. Nussbaum (Oxford: Oxford University Press, 2017), 9–11.

126. Kimmerer, *Braiding Sweetgrass*, 359.

Chapter 2. Anticipating Loss: On Naming

1. Ted Floyd, *Field Guide to the Birds of North America* (New York: Harper Collins, 2008), 1.

2. I take my lead here from Hannah Arendt, who writes that "the sheer naming of things, the creation of words, is the human way of *appropriating* and, as it were, disalienating the world into which, after all, each of us is born as a newcomer and a stranger." See Hannah Arendt, *The Life of the Mind* (San Diego: Harcourt, 1977), 100.

3. I first developed this argument in Joshua Trey Barnett, "Naming, Mourning, and the Work of Earthly Coexistence," *Environmental Communication* 13, no. 3 (2019):

287–99.

4. For a truly superb explanation of the violence inherent in naming, see Jacques Derrida, *The Animal That Therefore I Am* (New York: Fordham University Press, 2008), 1–51.
5. Elizabeth Kolbert, *The Sixth Extinction: An Unnatural History* (New York: Picador, 2014), 1.
6. Friedrich Nietzsche makes a similar point in his dazzling little essay on rhetoric. There, he contends that the human desire to name is coeval with our desire to control by way of "knowing." Importantly, though, Nietzsche also regards naming as a creative act by which we might remake our relationship with others and with the world. See Friedrich Nietzsche, "On Truth and Lying in a Non-Moral Sense," in *The Birth of Tragedy and Other Writings*, ed. Raymond Guess and Ronald Speirs (Cambridge: Cambridge University Press, 1999), 178–92.
7. Lynn White Jr., "The Historical Roots of Our Ecological Crisis," *Science* 155, no. 3767 (1967): 1205.
8. Neil Evernden, *The Social Creation of Nature* (Baltimore: The Johns Hopkins University Press, 1992), 132.
9. Michael Ohl, *The Art of Naming*, trans. Elisabeth Lauffer (Cambridge, MA: MIT Press, 2018), vii.
10. George R. Stewart, *Names on the Land: A Historical Account of Place-Naming in the United State*s (New York: New York Review of Books, 2008), 3.
11. Keith H. Basso, *Wisdom Sits in Places: Landscape and Language among the Western Apache* (Albuquerque: University of New Mexico Press, 1996), 76.
12. Thomas F. Thornton, *Being and Place among the Tlingit* (Seattle: University of Washington Press, 2008), 31.
13. Thomas Thornton addresses this example in detail. See Thornton, *Being and Place*, 68–115.
14. Winona LaDuke, *Recovering the Sacred: The Power of Naming and Claiming* (Chicago: Haymarket, 2016), 132.
15. What US Americans presently call the Wisconsin River, for instance, was first recorded by the French explorers Louis Joliett and Jacques Marquette as "Mescousing" or "Mesconsing." See Stewart, *Names on the Land*, 88.
16. The boarding schools established by the US Office of Indian Affairs at the end of the

nineteenth century were one of the government's primary mechanisms for disappearing Indigenous languages. As Roxane Dunbar-Ortiz notes, "Indigenous children were prohibited from speaking their mother tongues" and "were beaten for speaking their own languages." See Roxanne Dunbar-Ortiz, *An Indigenous Peoples' History of the United States* (Boston: Beacon, 2014), 151.

17. Winona LaDuke, *All Our Relations: Native Struggles for Land and Life* (Chicago: Haymarket, 2015), 130.
18. Quoted in Thornton, *Being and Place*, 73.
19. Robin Wall Kimmerer, *Braiding Sweetgrass: Indigenous Wisdom, Scientific Knowledge, and the Teachings of Plants* (Minneapolis: Milkweed, 2013), 208.
20. Lauret Savoy, *Trace: Memory, History, Race, and the American Landscape* (Berkeley, CA: Counterpoint, 2015), 75.
21. See Thom van Dooren, *Flight Ways: Life and Loss at the Edge of Extinction* (New York: Columbia University Press, 2014), 116–22.
22. Derrida, *The Animal*, 9.
23. Derrida, *The Animal*, 20.
24. See the end of Jacques Derrida, "Signature Event Context," in *Margins of Philosophy*, trans. Alan Bass (Chicago: University of Chicago Press, 1982), 307–30.
25. Joan Kirkby, "'Remembrance of the Future': Derrida on Mourning," *Social Semiotics* 16, no. 3 (2006): 464.
26. Jacques Derrida, "The Taste of Tears," in *The Work of Mourning*, ed. Pascale-Anne Brault and Michael Naas (Chicago: University of Chicago Press, 2001), 107.
27. Judith Butler, *Precarious Life: The Powers of Mourning and Violence* (London: Verso, 2004), 46.
28. Butler, *Precarious Life*, 37.
29. Butler, *Precarious Life*, 146.
30. Robert Macfarlane, *Landmarks* (New York: Penguin, 2015), 24.
31. See Joel Greenberg, *A Feathered River across the Sky: The Passenger Pigeon's Flight to Extinction* (New York: Bloomsbury, 2014), 185–89.
32. Robert M. Webster and Bruce Erickson, "The Last Word?," *Nature* 380 (1996): 386.
33. For the responses, see Elaine Andrews, "The Last Word," *Nature* 381 (1996): 272; David Craig, "The Last Word," *Nature* 381 (1996): 272; Mark Smith, "The Last Word,"

Nature 381 (1996): 272.

34. On the myth of the vanishing Indian, see Roxanne Dunbar-Ortiz and Dina Gilio-Whitaker, *"All the Real Indians Died Off" and 20 Other Myths about Native Americans* (Boston: Beacon, 2016), 7–13.
35. Dolly Jørgensen, "Endling, the Power of the Last in an Extinction-Prone World," *Environmental Philosophy* 14, no. 1 (2017): 134.
36. Quoted in Michelle Nijhuis, "What Do You Call the Last of a Species?," *New Yorker*, March 2, 2017, https://www.newyorker.com/.
37. Mike Shanahan, "Unhappy Endlings: What Tales of the Last Days of Extinction and Dying Species Can Bring to Our Own Story," *Under the Banyan* (blog), April 2, 2013, https://underthebanyan.blog/2013/04/02/unhappy-endlings-what-tales-of-the-last-days-of-extinct-and-dying-species-can-bring-to-our-own-story/.
38. Jacques Derrida, *The Politics of Friendship*, trans. George Collins (London: Verso, 1997), 14.
39. Eric Freedman, "Extinction Is Forever: A Quest for the Last Known Survivors," *Earth Island Journal*, Autumn 2011, https://www.earthisland.org/journal/index.php/magazine/entry/extinction_is_forever/.
40. Van Dooren, *Flight Ways*, 11.
41. Van Dooren, *Flight Ways*, 12.
42. Ed Yong, "The Last of Its Kind," *Atlantic*, July 2019, https://www.theatlantic.com/.
43. Not only endlings receive proper names. Individual animals are often named by the people they interact with. Consider, for instance, the case of "Cecil," an African lion who was killed by an American dentist on July 2, 2015. Lions and other "big game" are killed regularly; each death does not spark outcry and public attention, however. I would suggest that Cecil's name contributed significantly to spurring public expressions of rage and grief that followed from his loss. On the naming of Cecil, see Andrew Loveridge, *Lion Hearted: The Life and Death of Cecil and the Future of Africa's Iconic Cats* (New York: Regan Arts, 2018), 6–8.
44. "The IUCN Red List of Threatened Species," IUCN Red List, 2020, https://www.iucnredlist.org.
45. See Heather Ackroyd and Dan Harvey, *Seeing Red.. Overdrawn*, Ackroyd & Harvey, 2016, https://www.ackroydandharvey.com/seeing-red-overdrawn/.

46. Akiko Busch, *How to Disappear: Notes on Invisibility in a Time of Transparency* (New York: Penguin, 2019), 108.
47. Ackroyd and Harvey, *Seeing Red.. Overdrawn.*
48. Robert Pogue Harrison, *The Dominion of the Dead* (Chicago: University of Chicago Press, 2003), 138.
49. See "Designer of the Vietnam Veterans Memorial: Maya Lin," Vietnam Veterans Memorial Fund, 2020, https://www.vvmf.org/About-The-Wall/history-of-the-vietnam-veterans-memorial/Maya-Lin/.
50. *Seeing Red.. Overdrawn* differs, too, from the walls of names of extinct species in many natural history museums. For an overview of those displays, see Dolly Jørgensen, "Naming the Victims of the Sixth Mass Extinction," Penn Program in Environmental Humanitites Lab, posted April 15, 2020, YouTube video titled "Naming Extinction with Dolly Jørgensen," 19:25, https://www.youtube.com/watch?v=adWfrKYH_GM&feature=youtu.be.
51. Derrida, *The Politics of Friendship*, 14.
52. Quoted in Alison Flood, "Oxford Junior Dictionary's Replacement of 'Natural' Words with 21st-Century Terms Sparks Outcry," *Guardian*, January 13, 2015, https://www.theguardian.com/.
53. Stefan Fatsis, "Panic at the Dictionary," *New Yorker*, January 30, 2015, https://www.newyorker.com/books/.
54. Macfarlane, *Landmarks*, 3.
55. Macfarlane, *Landmarks*, 4.
56. Fatsis, "Panic at the Dictionary."
57. A PDF version of the Atwood et al. letter has long been available online at: http://www.naturemusicpoetry.com/uploads/2/9/3/8/29384149/letter_to_oup_final.pdf.
58. Robert Macfarlane and Jackie Morris, *The Lost Words: A Spell Book* (Toronto: House of Anansi, 2018), n.p.
59. Macfarlane, *Landmarks*, 24.
60. Robert Hass, "Ford," in *Home Ground: A Guide to the American Landscape*, ed. Barry Lopez and Debra Gwartney (San Antonio: Trinity University Press, 2006), 186.
61. Joy Williams, "Fosse," in *Home Ground: A Guide to the American Landscape*, ed. Barry Lopez and Debra Gwartney (San Antonio: Trinity University Press, 2006), 189.

62. Macfarlane, *Landmarks*, 39.

63. Macfarlane, *Landmarks*, 380.

64. Clive Hamilton, *Defiant Earth: The Fate of Humans in the Anthropocene* (Cambridge, UK: Polity, 2017), 17.

65. Bill McKibben, *Eaarth: Making a Life on a Tough New Planet* (New York: St. Martin's Press, 2010), 2.

66. See Jacques Derrida, "Différance," in *Margins of Philosophy*, trans. Alan Bass (Chicago: University of Chicago Press, 1982), 1–27.

67. Paul Huebener, *Nature's Broken Clocks: Reimagining Time in the Face of Environmental Crisis* (Regina, CA: University of Regina Press, 2020), 66.

68. Paul Crutzen and Eugene Stoermer, "The 'Anthropocene,'" *Global Change Newsletter* 41 (2000): 18.

69. The decision whether to formally accept "Anthropocene" as the name of a specific geological time period belongs not to individuals but to the scientific community. In particular, the International Union of Geological Sciences must ratify a proposal to so name our current geological epoch. In 2019, the Subcommittee on Quaternary Stratigraphy voted to formally recognize the Anthropocene as a "formal chrono-stratigraphic unit" with its origins in "one of the stratigraphic signals around the mid-twentieth century of the Common Era." For now, however, "Anthropocene" remains an informal designation. See "Working Group on the 'Anthropocene,'" Subcommission on Quaternary Stratigraphy, n.d., http://quaternary.stratigraphy.org/working-groups/anthropocene/.

70. Zoe Todd, "Indigenizing the Anthropocene," in *Art in the Anthropocene: Encounters among Aesthetics, Politics, Environments and Epistemologies*, ed. Heather Davis and Etienne Turpin (London: Open Humanities Press, 2015), 244.

71. See Andreas Malm and Alf Hornborg, "The Geology of Mankind? A Critique of the Anthropocene Narrative," *Anthropocene Review* 1, no. 1 (2014): 62–69.

72. See Sarah Jaquette Ray, *The Ecological Other: Environmental Exclusion in American Culture* (Tucson: University of Arizona Press, 2013), 21.

73. Eileen Crist, "On the Poverty of Our Nomenclature," *Environmental Humanities* 3, no. 1 (2013): 140.

74. See Franciszek Chwałczyk, "Around the Anthropocene in Eighty Names: Considering

the Urbanocene Proposition," *Sustainability* 12, no. 11 (2020): 23–29.

75. Jason W. Moore, "The Capitalocene, Part I: On the Nature and Origins of Our Ecological Crisis," *Journal of Peasant Studies* 44, no. 3 (2017): 594.
76. For instance, see Roy Scranton, *Learning to Die in the Anthropocene* (San Francisco: City Lights, 2015).
77. Nancy Menning, "Environmental Mourning and the Religious Imagination," in *Mourning Nature: Hope at the Heart of Ecological Loss and Grief*, ed. Ashlee Cunsolo and Karen Landman (Montreal: McGill–Queen's University Press, 2017), 40.
78. For a discussion of the losses associated with climate change mitigation, see Rosemary Randall, "Loss and Climate Change: The Cost of Parallel Narratives," *Ecopsychology* 1, no. 3 (2009): 118–29.

Chapter 3. Revealing Loss: On Archiving

1. Aldo Leopold, *A Sand County Almanac and Sketches Here and There* (New York: Oxford University Press, 1949), 32.
2. Quoted in Gaston R. Demarée and This Rutishauser, "From 'Periodical Observations' to 'Anthochronology' and 'Phenology': The Scientific Debate between Adolphe Quetelet and Charles Morren on the Origin of the Word 'Phenology,'" *International Journal of Biometeorology* 55, no. 6 (2011): 758.
3. John Durham Peters, *The Marvelous Clouds: Toward a Philosophy of Elemental Media* (Chicago: University of Chicago Press, 2015), 7.
4. Aldo Leopold and Sara Elizabeth Jones, "A Phenological Record for Sauk and Dane Counties, Wisconsin, 1935–1945," *Ecological Monographs* 17, no. 1 (1947): 83.
5. Mark L. Hineline, *Ground Truth: A Guide to Tracking Climate Change at Home* (Chicago: University of Chicago Press, 2018), 43.
6. See Thomas Richmond Kenote, "Indigenous Phenology: An Interdisciplinary Case Study on Indigenous Phenological Knowledge on the Menominee Nation Forest" (master's thesis, University of Minnesota, 2020), 39–47.
7. See Samantha Chisholm Hatfield et al., "Indian Time: Time, Seasonality, and Culture in Traditional Ecological Knowledge of Climate Change," *Ecological Processes* 7, no. 25

(2018): 7.

8. Karim-Aly S. Kassam et al., "Anticipating Climatic Variability: The Potential of Ecological Calendars," *Human Ecology* 46, no. 2 (2018): 250.

9. Kassam et al., "Anticipating Climatic Variability," 250.

10. Hatfield et al., "Indian Time," 5.

11. Hineline, *Ground Truth*, 45.

12. Thomas Jefferson, *The Papers of Thomas Jefferson, Volume 40: 4 March to 10 July 1803*, ed. Barbara B. Oberg (Princeton, NJ: Princeton University Press, 2013), 178.

13. See Leo Stoller, "A Note on Thoreau's Place in the History of Phenology," *Isis* 47, no. 2 (1956): 172–81; Sarah Dimick, "Disordered Environmental Time: Phenology, Climate Change, and Seasonal Form in the Work of Henry David Thoreau and Aldo Leopold," *ISLE* 25, no. 4 (2018): 700–721.

14. See Kristen Case, "Knowing as Neighboring: Approaching Thoreau's Kalendar," *J19: The Journal of Nineteenth Century Americanists* 2, no. 1 (2014): 107–29. In addition to their scholarship on Thoreau's Kalendar, Case has also digitized much of Thoreau's phenological data. See Kristen Case, "About the Kalendar," Thoreau's Kalendar, n.d., https://thoreauskalendar.org/about.html.

15. Leopold and Jones, "A Phenological Record."

16. Rachel Carson, *Silent Spring* (Boston: Houghton Mifflin, 1962).

17. Aldo Leopold, "The Round River," in *A Sand County Almanac with Essays on Conservation from Round River* (New York: Ballantine, 1966), 197.

18. Eccles. 1:18 (New King James Version).

19. In his field-defining textbook, *Game Management* (originally published in 1933), Leopold advises students that "phenological tables" are useful tools in making sense of "game phenomena." Aldo Leopold, *Game Management* (Madison: University of Wisconsin Press, 1986), 165.

20. Leopold, *A Sand County Almanac*, vii–viii.

21. A brief account of the property can be found at "The Shack," The Aldo Leopold Foundation, 2020, https://www.aldoleopold.org/visit/the-shack/.

22. Curt Meine, *Aldo Leopold: His Life and Work*, 2nd ed. (Madison: University of Wisconsin Press, 2010), 341.

23. "Who We Are, What We Do," Aldo Leopold Foundation, 2020, https://www.

aldoleopold.org/about/our-work/.

24. Leopold, *A Sand County Almanac*, 44.
25. Leopold, *A Sand County Almanac*, 45.
26. Leopold, *A Sand County Almanac*, 46.
27. Leopold, *A Sand County Almanac*, 46.
28. Leopold, *A Sand County Almanac*, 48.
29. Leopold, *A Sand County Almanac*, 46.
30. See Timothy Morton, *The Ecological Thought* (Cambridge, MA: Harvard University Press, 2010).
31. See David Wallace-Wells, *The Uninhabitable Earth: Life after Warming* (New York: Tim Duggan, 2019).
32. Leopold, *A Sand County Almanac*, 50.
33. Leopold's insights are confirmed in Rich Wynia, "Plant Fact Sheet for Compassplant (*Silphium laciniatum* L.)" (US Department of Agriculture Natural Resources Conservation Service, June 2009), https://plants.usda.gov/DocumentLibrary/factsheet/pdf/fs_sila3.pdf.
34. Leopold, *A Sand County Almanac*, 50.
35. On the history of the extermination of the American bison, see Winona LaDuke, *All Our Relations: Native Struggles for Land and Life* (Chicago: Haymarket, 2015), 139–62.
36. "*Silphium Laciniatum* L., Compass Plant," USDA Natural Resources Conservation Service, n.d., https://plants.usda.gov/core/profile?symbol=SILA3.
37. Sarah D. Wright and Nina Leopold Bradley, "Thinking Like a Flower: Phenology and Climate Change at the Leopold Shack," in *The Vanishing Present: Wisconsin's Changing Lands, Waters, and Wildlife*, ed. Donald M. Waller and Thomas P. Rooney (Chicago: University of Chicago Press, 2008), 46.
38. See Elizabeth R. Ellwood et al., "Record-Breaking Early Flowering in the Eastern United States," *PLOS One* 8, no. 1 (2013): 1–9.
39. Wright and Bradley, "Thinking Like a Flower," 51.
40. Hineline, *Ground Truth*, 56.
41. Dimick, "Disordered Environmental Time," 712.
42. Dimick, "Disordered Environmental Time," 713.

43. Leopold and Jones, "A Phenological Record," 83, emphasis mine.
44. "Great Acceleration," International Geosphere-Biosphere Programme, 2015, http://www.igbp.net/globalchange/greatacceleration.4.1b8ae20512db692f2a680001630.html.
45. Nina L. Bradley et al., "Phenological Changes Reflect Climate Change in Wisconsin," *Proceedings of the National Academy of Science of the United States of America* 96, no. 17 (1999): 9703.
46. See Ellwood et al., "Record-Breaking Early Flowering."
47. Thom van Dooren, *Flight Ways: Life and Loss at the Edge of Extinction* (New York: Columbia University Press, 2014), 140.
48. Hineline, *Ground Truth*, 99.
49. Leopold, *A Sand County Almanac*, 44.
50. Quoted in Akiko Busch, *The Incidental Steward: Reflections on Citizen Science* (New Haven, CT: Yale University Press, 2013), 119.
51. Leopold, *A Sand County Almanac*, 214.
52. Busch, *The Incidental Steward*, 181.
53. Busch, *The Incidental Steward*, 201.
54. Written records are not, however, the only means by which ecological knowledge is transmitted from one generation to the next. Such knowledge can be transmitted orally as well. Nonetheless, written records enable us to speak with a degree of precision that oral utterances may not. While it would be interesting to consider how an emphasis on such precision in phenology may obscure other important insights about seasonality, such as the ways in which some people view seasons are more about relationships than about calendar dates, this is not a line of thought I take up here.
55. Meine, *Aldo Leopold*, 472.
56. "Leopold Shack Journal, 1946–1948," box 003, folder 003, Aldo Leopold Papers, University of Wisconsin-Madison Libraries, http://digital.library.wisc.edu/1711.dl/AldoLeopold.
57. Donna Haraway, *When Species Meet* (Minneapolis: University of Minnesota Press, 2008), 36.
58. Richard B. Primack, *Walden Warming: Climate Changes Comes to Thoreau's Woods* (Chicago: University of Chicago Press, 2014), 230.

59. Busch, *The Incidental Steward*, 8.
60. Hineline, *Ground Truth*, 53.
61. Jacques Derrida, *Archive Fever: A Freudian Impression*, trans. Eric Prenowitz (Chicago: University of Chicago Press, 1995), 19.
62. Ann Cvetkovich, *An Archive of Feelings: Trauma, Sexuality, and Lesbian Public Cultures* (Durham, NC: Duke University Press, 2003), 210–11.
63. See, for instance, Terry Cook and Joan M. Schwartz, "Archives, Records, and Power: From (Postmodern) Theory to (Archival) Performance," *Archival Science* 2, no. 3 (2002): 171–85.
64. Pamela VanHaitsma, "Digital LGBTQ Archives as Sites of Public Memory and Pedagogy," *Rhetoric & Public Affairs* 22, no. 2 (2019): 253–54.
65. Barbara Biesecker, "Of Historicity, Rhetoric: The Archive as Scene of Invention," *Rhetoric & Public Affairs* 9, no. 1 (2006): 130.
66. Charles E. Morris III, "Archival Queer," *Rhetoric & Public Affairs* 9, no. 1 (2006): 147.
67. These stories are shaped, too, by which records are permitted a place in an archives. As Michelle Caswell points out, "archivists do not keep everything." Rather, "they make appraisal decisions" about what to include, what to dispose of, and what to send elsewhere. Michelle Caswell, "'The Archive' Is Not An Archives: Acknowledging the Intellectual Contributions of Archival Studies," *Reconstruction* 16, no. 1 (2016): para. 16, https://escholarship.org/uc/item/7bn4v1fk.
68. Hineline, *Ground Truth*, 52–53.
69. Biesecker, "Of Historicity, Rhetoric," 124.
70. See "The Aldo Leopold Archives," University of Wisconsin–Madison Libraries, https://uwdc.library.wisc.edu/collections/aldoleopold/.
71. Florian Heigl et al., "Toward an International Definition of Citizen Science," *Proceedings of the National Academy of Science of the United States of America* 116, no. 17 (2019): 8089–92.
72. Busch, *The Incidental Steward*, 15.
73. Primack, *Walden Warming*, 54.
74. Derrida, *Archive Fever*, 17.
75. On the notion of "participation" in contemporary archives, see Ekaterina Haskins, "Between Archive and Participation: Public Memory in a Digital Age," *Rhetoric Society*

Quarterly 37, no. 4 (2007): 401–22.

76. "15 Million Records Submitted to the National Phenology Database," USA National Phenology Network, December 17, 2018, https://www.usanpn.org/node/35012.

77. For instance, see Adam G. Hart et al., "Testing the Potential of Twitter Mining Methods for Data Acquisition: Evaluating Novel Opportunities for Ecological Research in Multiple Taxa," *Methods in Ecology and Evolution* 9, no. 11 (2018): 2194–2205.

78. Joshua Trey Barnett, "Irrational Hope, Phenological Writing, and the Prospects of Earthly Coexistence," *Communication and Critical/Cultural Studies* 16, no. 4 (2019): 390.

79. Richard J. Hopp, "Plant Phenology Observation Networks," in *Phenology and Seasonality Modeling*, ed. Helmut Lieth (New York: Springer, 1974), 29.

80. Leopold and Jones, "A Phenological Record," 85.

81. Hineline, *Ground Truth*, 57–59.

82. "Volunteer Scientists," USA National Phenology Network, https://www.usanpn.org/partner/volunteer-scientists.

83. Scott Slovic, *Seeking Awareness in American Nature Writing: Henry Thoreau, Annie Dillard, Edward Abbey, Wendell Berry, Barry Lopez* (Salt Lake City: University of Utah Press, 1992), 4.

84. "Learn How to Observe," USA National Phenology Network, https://www.usanpn.org/nn/guidelines.

85. "Peer-Reviewed Results," USA National Phenology Network, https://www.usanpn.org/pubs/results.

86. Lori Petrauski et al., "Changes in Flowering Phenology of *Cardamine concatenata* and *Erythronium americanum* over 111 Years in the Central Appalachians," *Plant Ecology* 220, no. 9 (2019): 826.

87. Primack, *Walden Warming*, 233.

88. "Status of Spring," USA National Phenology Network, https://www.usanpn.org/news/spring.

89. Theresa Crimmins, associate director of USA-NPN, interviewed by Jim Cantore and Stephanie Abrams, on The Weather Channel, aired January 23, 2020, YouTube video, titled "USA-NPN on The Weather Channel January 23, 2020" 3:09, https://www.

youtube.com/watch?v=NMzLSvdppjQ.

90. This is perhaps nowhere more evident than in Richard Primack's study of the effects of climate change on Thoreau's beloved Walden. So thoroughly have conditions in Thoreau's woods changed in the 160 years since he made his own observations that, Primack contends, the naturalist might not recognize the place today. Primack, *Walden Warming*.

91. Cvetkovich, *An Archive of Feelings*, 211.

Chapter 4. Imagining Loss: On Making Visible

1. Gregory J. Nowacki and Marc D. Abrams, "Forest Composition, Structure, and Disturbance History of the Alan Seeger Natural Area, Huntington County, Pennsylvania," *Bulletin of the Torrey Botanical Club* 121, no. 3 (1994): 277.

2. David Orwig, "An Iconic Species," in *Hemlock: A Forest Giant on the Edge*, ed. David R. Foster (New Haven, CT: Yale University Press, 2014), 11.

3. I came across the phrase "forest giant" in the title of David R. Foster's edited collection on eastern hemlocks, from which much of my knowledge of the species has been derived. See David R. Foster, ed., *Hemlock: A Forest Giant on the Edge* (New Haven, CT: Yale University Press, 2014).

4. Aristotle, "Rhetoric," in *The Complete Works of Aristotle*, ed. Jonathan Barnes, trans. W. Rhys Roberts, vol. 2 (Princeton, NJ: Princeton University Press, 1984), 2202.

5. Aristotle, "Rhetoric," 2194.

6. See Aristotle, "Rhetoric," 2159–61.

7. Michele Kennerly, "Getting Carried Away: How Rhetorical Transport Gets Judgment Going," *Rhetoric Society Quarterly* 40, no. 3 (2010): 269.

8. This view is pervasive. As Richard Vatz contends, for instance, "This may be the *sine qua non* of rhetoric: the art of linguistically or symbolically creating salience." Chris Ingraham puts it still more simply: "rhetoric makes things matter in influential ways." Richard Vatz, "The Myth of the Rhetorical Situation," *Philosophy & Rhetoric* 6, no. 3 (1973): 160; Chris Ingraham, *Gestures of Concern* (Durham, NC: Duke University Press, 2020), 185.

9. Kevin Michael DeLuca, *Image Politics: The New Rhetoric of Environmental Activism* (New York: Guilford, 1999), 161; the story is recounted in more detail in Julia Butterfly Hill, *The Legacy of Luna: The Story of a Tree, a Woman, and the Struggle to Save the Redwoods* (San Francisco: Harper Collins, 2001).
10. Eliot Porter, *The Place No One Knew: Glen Canyon on the Colorado* (San Francisco: Sierra Club, 1963).
11. Jeff Orlowski, *Chasing Ice* (New York: Submarine Deluxe, 2012), documentary, 75 min.
12. "Survival by Degrees: 389 Bird Species on the Brink," Audubon Society, 2019, https://www.audubon.org/climate/survivalbydegrees.
13. Lester C. Olson, Cara A. Finnegan, and Diane S. Hope, "Visual Rhetoric in Communication: Continuing Questions and Contemporary Issues," in *Visual Rhetoric: A Reader in Communication and American Culture*, ed. Lester C. Olson, Cara A. Finnegan, and Diane S. Hope (Los Angeles: SAGE, 2008), 7.
14. As Kevin Michael DeLuca has noted, however, even scholars of visual rhetoric express a tendency to subordinate images to words. See Kevin Michael DeLuca, "The Speed of Immanent Images: The Dangers of Reading Photographs," in *Visual Communication: Perception, Rhetoric, and Technology*, ed. Diane S. Hope (Cresskill, NJ: Hampton, 2006), 79–90.
15. For instance, see Olson, Finnegan, and Hope, "Visual Rhetoric," 4–8; Cara Finnegan, "Communication as Vision," in *Communication As . . . Perspectives on Theory*, ed. Gregory J. Shepherd, Jeffrey St. John, and Ted Striphas (Thousand Oaks, CA: SAGE, 2006), 60–66.
16. I develop this point in Joshua Trey Barnett, "Toxic Portraits: Resisting Multiple Invisibilities in the Environmental Justice Movement," *Quarterly Journal of Speech* 101, no. 2 (2015): 405–25.
17. Jennifer Peeples, "Toxic Sublime: Imaging Contaminated Landscapes," *Environmental Communication* 5, no. 4 (2011): 373–92.
18. On this dynamic, see William Cronon, "The Trouble with Wilderness: Or, Getting Back to the Wrong Nature," in *Uncommon Ground: Rethinking the Human Place in Nature*, ed. William Cronon (New York: W. W. Norton, 1995), 80–90.
19. Davi Johnson, "Martin Luther King Jr.'s 1963 Birmingham Campaign as Image Event," *Rhetoric & Public Affairs* 10, no. 1 (2007): 2.

20. Johnson, "Martin Luther King," 3.
21. On this point, one cannot help but recall Michel Foucault's analysis of public torture and private discipline. See Michel Foucault, "The Body of the Condemned," in *Discipline and Punish: The Birth of the Prison*, trans. Alan Sheridan (New York: Vintage, 1995), 3–31.
22. Joshua Trey Barnett, "Grievable Water: Mourning the Animas River," in *Water, Rhetoric, and Social Justice: A Critical Confluence*, ed. Casey Schmitt, Christopher S. Thomas, and Theresa Castor (Lanham, MD: Lexington, 2020), 280.
23. See Joshua Trey Barnett, "Naming, Mourning, and the Work of Earthly Coexistence," *Environmental Communication* 13, no. 3 (2019): 287–99.
24. "The Passenger Pigeon," Smithsonian Institution, 2001, https://www.si.edu/spotlight/passenger-pigeon.
25. See Joel Greenberg, *A Feathered River across the Sky: The Passenger Pigeon's Flight to Extinction* (New York: Bloomsbury, 2014), 185–89.
26. See Thom van Dooren, *Flight Ways: Life and Loss at the Edge of Extinction* (New York: Columbia University Press, 2014), 11.
27. Greenberg, *A Feathered River*, 1.
28. John James Audubon, *Ornithological Biography, or An Account of the Habits of the Birds of the United States of America* (Philadelphia, 1831), 320.
29. Audubon, *Ornithological Biography*, 323.
30. Greenberg, *A Feathered River*, 69.
31. A copy of Bennett's sketch can be found in the plate section, between pages 162 and 163, of Greenberg's book. Greenberg, *A Feathered River*.
32. Audubon, *Ornithological Biography*, 325.
33. Gertrude M. Scott, "Society Meets in May at Madison," *The Passenger Pigeon* 9, no. 2 (1947): 58–59.
34. See Walter Edwin Scott, ed., *Silent Wings: A Memorial to the Passenger Pigeon* (Madison: Wisconsin Society for Ornithology, 1947).
35. Aldo Leopold, "On a Monument to the Pigeon," in *Silent Wings: A Memorial to the Passenger Pigeon*, ed. Walter Edwin Scott (Madison: Wisconsin Society for Ornithology, 1947), 3.
36. I recount this experience in more detail in Joshua Trey Barnett, "Vigilant Mourning

and the Future of Earthly Coexistence," in *Communicating in the Anthropocene: Intimate Relations*, ed. Alexa Dare and C. Vail Fletcher (Lanham, MD: Lexington, 2021), 13–33.

37. Wyalusing State Park is home to nearly seventy extant Indian mounds; at one point, archaeologists identified nearly two hundred mounds within the park's borders. See "Wyalusing State Park," Wisconsin First Nations, https://wisconsinfirstnations.org/wyalusing-state-park/.
38. In 1949, Leopold could write: "Men still live who, in their youth, remember pigeons. Trees still live who, in their youth, were shaken by a living wind. But a decade hence only the oldest oaks will remember, and at long last only the hills will know." See Aldo Leopold, *A Sand County Almanac and Sketches Here and There* (New York: Oxford University Press, 1949), 109.
39. Carole Blair, Greg Dickinson, and Brian L. Ott, "Rhetoric/Memory/Place," in *Places of Public Memory: The Rhetoric of Museums and Memorials*, ed. Greg Dickinson, Carole Blair, and Brian L. Ott (Tuscaloosa: University of Alabama Press, 2010), 13.
40. On the issue of contestation, see Thomas R. Dunn, "Remembering 'A Great Fag': Visualizing Public Memory and the Construction of Queer Space," *Quarterly Journal of Speech* 97, no. 4 (2011): 435–60.
41. Jennifer K. Ladino, *Memorials Matter: Emotion, Environment, and Public Memory at American Historical Sites* (Reno: University of Nevada Press, 2019), xix.
42. Ladino, *Memorials Matter*, xiii.
43. Marilyn DeLaure, "Performing Loss: Sonic Rhetoric in Maya Lin's *What Is Missing?*," *Liminalities: A Journal of Performance Studies* 16, no. 1 (2020): 9.
44. Leopold, *A Sand County Almanac*, 111.
45. Helen Whale and Franklin Ginn, "In the Absence of Sparrows," in *Mourning Nature: Hope at the Heart of Ecological Loss and Grief*, ed. Ashlee Cunsolo and Karen Landman (Montreal: McGill–Queen's University Press, 2017), 99.
46. As Ladino notes, memorials are often sites of "unruly affects" and "affective dissonance," as when we "experience more than one *feeling* at the same time." See Ladino, *Memorials Matter*, 22.
47. Leopold, *A Sand County Almanac*, 112.
48. Leopold, "On a Monument," 4.

49. Barnett, "Vigilant Mourning."

50. Robert A. Birmingham and Amy L. Rosebrough, *Indian Mounds of Wisconsin*, 2nd ed. (Madison: University of Wisconsin Press, 2017), 3.

51. John A. Cross, "Native American Landscapes in the Eastern United States," in *Ethnic Landscapes of America* (Cham, Switzerland: Springer Nature, 2017), 22.

52. Since I visited Wyalusing State Park in 2019, the sign near the Indian mounds has been replaced. The new sign reads: "Sentinel Ridge Mound Group. The mound group you see before you, the largest in the Park, includes numerous conical and linear forms, as well as one bear mound. The meaning of the effigy forms is not clear, but they likely reflected shared group identities. . . . Indian peoples have been respectfully burying their dead ever since they settled in these lands many thousands of years ago. During what is called the Woodland Period (ca. 500 BC to AD 1250), individuals were buried in earthen mounds, typically conical (round) in form. Later in this period, linear and animal-shaped effigy mounds were commonly built as well. Wisconsin has, by far, the highest concentration of such mounds in the world. Sadly, however, many thousands of mounds have been destroyed in past times as a result of farming and other development related activities." I wish to thank Joshua Kozelka, a naturalist at Wyalusing State Park, for sharing photographs of the new signs with me.

53. See Elizabeth Dohms-Harter, "'A Sculpted Land': Wisconsin's Effigy Mounds Connected People to Spirits, Humanity," *Wisconsin Public Radio* (blog), September 2, 2020, https://www.wpr.org/.

54. Rhetorical critic Casey Schmitt argues that Indian mounds are often appropriated for Western histories in modern times. Although mounds may be preserved in Western institutions, including state parks, the ways in which visitors are invited to interact with the mounds regularly obscures the violence inherent in settler colonialism and frames mound builders and their descendants in the past tense. See Casey R. Schmitt, "Contours of the Land: Place-as-Rhetoric and Native American Effigy Mounds," *Western Journal of Communication* 79, no. 3 (2015): 308.

55. Schmitt, "Contours of the Land," 323.

56. Elke Weber, Irena Bauman, and Olafur Eliasson, "Can Art Inspire Climate Change Action? An Ice Installation Aims to Do Just That," *Guardian*, October 23, 2014, https://www.theguardian.com/.

57. Nina Azzarello, "Olafur Eliasson Moves 100 Tonnes of Ice to Copenhagen to

Visualize Climate Change," *Designboom* (blog), October 24, 2014, https://www.designboom.com/art/olafur-eliasson-ice-watch-project-climate-change-greenland-copenhagen-10–24–2014/.

58. See *Ice Watch, 2014*, Olafur Eliasson, https://olafureliasson.net/archive/artwork/WEK109190/ice-watch.
59. I am grateful to Júlia Frate Bolliger of Studio Olafur Eliasson for sharing a copy of the artist's booklet with me. See *Ice Watch* (Studio Olafur Eliasson, 2015).
60. Among others, philosopher Robert Kirkman contends that our inability to perceive climatic changes (and their effects on our own lives) is the principle impediment to mitigating global warming. See Robert Kirkman, "A Little Knowledge of Dangerous Things: Human Vulnerability in a Changing Climate," in *Merleau-Ponty and Environmental Philosophy: Dwelling on the Landscape of Thought*, ed. Suzanne L. Cataldi and William S. Hamrick (Albany: State University of New York Press, 2007), 19–35.
61. See Jennifer Marlon et al., "Yale Climate Opinion Maps 2019," Yale Program on Climate Change Communication, September 17, 2019, https://climatecommunication.yale.edu/visualizations-data/ycom-us/.
62. Liz Creel, "Ripple Effects: Population and Coastal Regions" (Washington, DC: Population Review Bureau, September 25, 2003), https://www.prb.org/rippleeffectspopulationandcoastalregions/.
63. Whale and Ginn, "In the Absence," 92.
64. Walter Benjamin, "The Work of Art in the Age of Its Technological Reproducibility," in *The Work of Art in the Age of Its Technological Reproducibility, and Other Writings on Media*, ed. Michael W. Jennings, Brigid Doherty, and Thomas Y. Levin, trans. Edmund Jephcott, Rodney Livingstone, and Howard Eiland (Cambridge, MA: Harvard University Press, 2008), 23.
65. *Ice Watch*, 4.
66. These images are displayed at *Ice Watch, 2014*.
67. Tim Jonze, "Icebergs Ahead! Olafur Eliasson Brings the Frozen Fjord to Britain," *Guardian*, December 11, 2018, https://www.theguardian.com/.
68. Roy Scranton, *Learning to Die in the Anthropocene* (San Francisco: City Lights, 2015), 23.
69. Core Writing Team, Rajendra K. Pachauri, and Leo A. Meyer, "Climate Change 2014: Synthesis Report" (Geneva: Intergovernmental Panel on Climate Change, 2014),

https://www.ipcc.ch/site/assets/uploads/2018/02/SYR_AR5_FINAL_full.pdf.

70. The so-called "ecomodernists" suggest that we can retool our way into a "good Anthropocene." See John Asafu-Adjaye et al., "An Ecomodernist Manifesto," 2015, https://static1.squarespace.com/static/5515d9f9e4b04d5c3198b7bb/t/552d37bbe4b07a7dd69fcdbb/1429026747046/An+Ecomodernist+Manifesto.pdf.

71. For some examples, see Soren C. Larsen and Jay T. Johnson, *Being Together in Place: Indigenous Coexistence in a More Than Human World* (Minneapolis: University of Minnesota Press, 2017).

72. See Robert Graves and Didier Madoc-Jones, "Images," Postcards from the Future, http://www.postcardsfromthefuture.com/postcard_images/.

73. Robert Graves and Didier Madoc-Jones, "The Project," Postcards from the Future, http://www.postcardsfromthefuture.com/the-project/.

74. See "London Futures / Metro and Museum of London Competition," Flickr, 2010, https://www.flickr.com/groups/londonfuturesmetro/.

75. Graves and Madoc-Jones, "Images."

76. George Marshall, "Fantasy Images of Climate Migration Will Fuel Existing Prejudices," *Guardian*, October 27, 2010, https://www.theguardian.com/.

77. Graves and Madoc-Jones, "Images."

78. Ingraham, *Gestures of Concern*, 1.

79. In this sense, they compose what Haley Schneider calls a "public future." Haley Schneider, "Public Futures: The Rhetoric of Sustainability and Survival in Philadelphia's Environmental Initiatives" (lecture, Center for Democratic Deliberation Lecture Series, Pennsylvania State University, University Park, PA, March 24, 2021).

80. Scranton, *Learning to Die*, 38.

Epilogue. Caring Rhetorics: Meditations

1. Oddur Sigurðsson, a glaciologist for the Icelandic government, declared Okjökull "dead" in 2014. See Jane C. Hu, "How Can You Tell When a Glacier Is Dead?," *Slate*, July 24, 2019, https://slate.com/.

2. Joshua Stephens, "Okjökull Remembered," NASA Visible Earth, August 9, 2019,

https://visibleearth.nasa.gov/images/145439/okjokull-remembered.

3. Laura M. Holson, "Iceland Mourns Loss of a Glacier by Posting a Warning about Climate Change," *New York Times*, August 19, 2019, https://www.nytimes.com/.

4. For a brief history of the project, see Amy McCaig, "Saying Goodbye to Okjökull: Howe Reflects on Glacier Memorial Ceremony," *Rice University News and Media Relations* (blog), September 10, 2019, https://news.rice.edu/2019/09/10/saying-goodbye-to-okjokull-howe-reflects-on-glacier-memorial-ceremony/.

5. *Oxford English Dictionary Online*, s.v. "mourn (*v.1*)," https://www.oed.com/.

6. Mattha Busby, "Extinction Rebellion Activists Stage Die-In Protests across Globe," *Guardian*, April 27, 2019, https://www.theguardian.com/.

7. In staging die-ins, Extinction Rebellion activists participate in a long tradition of embodied ecological argument, a tradition in which vulnerable bodies supplement good reasons and rational arguments. See Kevin Michael DeLuca, "Unruly Arguments: The Body Rhetoric of Earth First!, ACT UP, and Queer Nation," *Argumentation and Advocacy* 36, no. 1 (1999): 9–21; Julia Butterfly Hill, *The Legacy of Luna: The Story of a Tree, a Woman, and the Struggle to Save the Redwoods* (San Francisco: Harper Collins, 2001).

8. "About Us," Extinction Rebellion, https://rebellion.global/about-us/.

9. The literature on care and care ethics is expansive. Many theorists of care view caring through a humanistic framework; however, from the early days of care ethics to the present, caring for other animals, plants, and the environment has been a part of the dialogue. For a sampling of such work, see María Puig de la Bellacasa, *Matters of Care: Speculative Ethics in More Than Human Worlds* (Minneapolis: University of Minnesota Press, 2017); The Care Collective, *The Care Manifesto: The Politics of Interdependence* (London: Verso, 2020); Deane Curtin, "Toward an Ecological Ethic of Care," *Hypatia* 6, no. 1 (1991): 60–74; Thom van Dooren, *Flight Ways: Life and Loss at the Edge of Extinction* (New York: Columbia University Press, 2014); Virginia Held, *The Ethics of Care: Personal, Political, and Global* (Oxford: Oxford University Press, 2006); Hil Malatino, *Trans Care* (Minneapolis: University of Minnesota Press, 2020); Deborah Bird Rose, *Wild Dog Dreaming: Love and Extinction* (Charlottesville: University of Virginia Press, 2011); Joan Tronto, *Moral Boundaries: A Political Argument for an Ethic of Care* (New York: Routledge, 1993).

10. An ethic of ecological and cultural care is at the beating heart of much writing by and about Indigenous peoples' relations with more-than-human others and with the land. For instance, see Nick Estes, *Our History Is the Future: Standing Rock versus the Dakota Access Pipeline, and the Long Tradition of Indigenous Resistance* (London: Verso, 2019); Dina Gilio-Whitaker, *As Long as Grass Grows: The Indigenous Fight for Environmental Justice: From Colonization to Standing Rock* (Boston: Beacon, 2019); Soren C. Larsen and Jay T. Johnson, *Being Together in Place: Indigenous Coexistence in a More Than Human World* (Minneapolis: University of Minnesota Press, 2017); Robin Wall Kimmerer, *Gathering Moss: A Natural and Cultural History of Mosses* (Corvalis: Oregon State University Press, 2003); Robin Wall Kimmerer, *Braiding Sweetgrass: Indigenous Wisdom, Scientific Knowledge, and the Teachings of Plants* (Minneapolis: Milkweed, 2013); Winona LaDuke, *All Our Relations: Native Struggles for Land and Life* (Chicago: Haymarket, 2015); Winona LaDuke, *Recovering the Sacred: The Power of Naming and Claiming* (Chicago: Haymarket, 2016); Kyle Powys Whyte, "Against Crisis Epistemology," in *Routledge Handbook of Critical Indigenous Studies*, ed. Brendan Hokowhitu et al. (London: Routledge, 2020), 52–64.
11. Robert Cox, "Nature's 'Crisis Disciplines': Does Environmental Communication Have an Ethical Duty?," *Environmental Communication* 1, no. 1 (2007): 10.
12. For instance, see Elizabeth Brunner, *Environmental Activism, Social Media, and Protest in China: Becoming Activists over Wild Public Screens* (Lanham, MD: Lexington, 2019); Jennifer Clary-Lemon, *Planting the Anthropocene: Rhetorics of Natureculture* (Logan: Utah State University Press, 2019); Kevin Michael DeLuca, *Image Politics: The New Rhetoric of Environmental Activism* (New York: Guilford, 1999); S. Marek Muller, *Impersonating Animals: Rhetoric, Ecofeminism, and Animal Rights Law* (East Lansing: Michigan State University Press, 2020); Phaedra C. Pezzullo, *Toxic Tourism: Rhetorics of Pollution, Travel, and Environmental Justice* (Tuscaloosa: University of Alabama Press, 2007).
13. Phaedra C. Pezzullo, "Environment," in *Oxford Research Encyclopedia of Communication*, ed. Jon F. Nussbaum (Oxford: Oxford University Press, 2017), 11.
14. The connection between ecological rhetoric and care also finds expression in Phaedra C. Pezzullo and Robert Cox, *Environmental Communication and the Public Sphere*, 5th ed. (Los Angeles: SAGE, 2018), 18; Justine Wells et al., "Introduction: Rhetoric's

Ecologies," in *Tracing Rhetoric and Material Life: Ecological Approaches*, ed. Bridie McGreavy et al. (London: Palgrave Macmillan, 2018), 1–36; and Danielle Endres, "Environmental Criticism," *Western Journal of Communication* 84, no. 3 (2020): 314–31.

15. Estes, *Our History*, 15.
16. Estes, *Our History*, 9.
17. Estes, *Our History*, 51.
18. For one especially significant set of takes on the scope of rhetoric, see Lloyd F. Bitzer and Edwin Black, eds., *The Prospect of Rhetoric: Report of the National Development Project* (Englewood Cliff, NJ: Prentice-Hall, 1971).
19. Aristotle, "Rhetoric," in *The Complete Works of Aristotle*, ed. Jonathan Barnes, trans. W. Rhys Roberts, vol. 2 (Princeton, NJ: Princeton University Press, 1984), 2155.
20. For a useful critique of the instrumental approach to assessing rhetoric, see David Zarefsky, "Presidential Rhetoric and the Power of Definition," *Presidential Studies Quarterly* 34, no. 3 (2004): 608–11.
21. Among others, Thomas Farrell and Chris Ingraham adopt a similar approach. See Thomas B. Farrell, "The Weight of Rhetoric: Studies in Cultural Delirium," *Philosophy & Rhetoric* 41, no. 4 (2008): 467–87; Chris Ingraham, *Gestures of Concern* (Durham, NC: Duke University Press, 2020).
22. Emily Laurens, "Remembrance Day for Lost Species," *Emily Laurens* (blog), November 28, 2014, https://emilylaurens.wordpress.com/2014/11/28/remembrance-day-for-lost-species/.
23. I consider Laurens's work in Joshua Trey Barnett, "Vigilant Mourning and the Future of Earthly Coexistence," in *Communicating in the Anthropocene: Intimate Relations*, ed. Alexa Dare and C. Vail Fletcher (Lanham, MD: Lexington, 2021), 13–33.
24. Laurens, "Remembrance Day for Lost Species."
25. Here, I follow Kathryn Yusoff's suggestion that by making violence visible we take a step toward engendering less violent worlds. See Kathryn Yusoff, "Aesthetics of Loss: Biodiversity, Banal Violence, and Biotic Subjects," *Transactions of the Institute of British Geographers* 37, no. 4 (2011): 578–92.
26. In relation to rhetorical studies, I make this argument most explicitly in Joshua Trey Barnett, "Rhetoric for Earthly Coexistence: Imagining an Ecocentric Rhetoric,"

Rhetoric & Public Affairs 24, no. 1–2 (2021): 365–78.

27. On shifting away from the "sky is falling" pedagogical approach of "critique-exposé," see Janell Watson and Jane Bennett, "Eco-sensibilities: An Interview with Jane Bennett," *Minnesota Review* 81 (2013): 151–52.

28. Sigmund Freud, "Mourning and Melancholia," in *Collected Papers: Volume IV*, ed. Joan Riviere (London: Hogarth, 1953); Van Dooren, *Flight Ways*; Ashlee Cunsolo and Karen Landman, "To Mourn Beyond the Human," in *Mourning Nature: Hope at the Heart of Ecological Loss and Grief*, ed. Ashlee Cunsolo and Karen Landman (Montreal: McGill–Queen's University Press, 2017), 3–26.

29. Details and instructions for this assignment can be found at Joshua Trey Barnett, "Eulogy Essay and Reading," UC-CSU NXTerra, https://www.nxterra.orfaleacenter.ucsb.edu/wp-content/uploads/2020/09/Assignment-Climate-Change-Emotions-Eulogy-Essay-and-Presentation-Joshua-Barnett.pdf.

30. On emotional intelligence, see Daniel Goleman, *Emotional Intelligence: Why It Can Matter More Than IQ* (New York: Bantam, 1995); Sarah Jaquette Ray, "Coming of Age at the End of the World: The Affective Arc of Undergraduate Environmental Studies Curricula," in *Affective Ecocriticism: Emotion, Embodiment, Environment*, ed. Kyle Bladow and Jennifer Ladino (Lincoln: University of Nebraska Press, 2018), 299–319; Sarah Jaquette Ray, *A Field Guide to Climate Anxiety: How to Keep Your Cool on a Warming Planet* (Oakland: University of California Press, 2020).

31. See William Cronon, "'Only Connect . . . ': The Goals of a Liberal Education," *American Scholar* 67, no. 4 (1998): 73–80.

32. For more details and images of the exhibit, see "Hemlock Hospice," Harvard Forest, https://harvardforest.fas.harvard.edu/hemlock-hospice.

33. David R. Foster, "Preface," in *Hemlock: A Forest Giant on the Edge*, ed. David R. Foster (New Haven, CT: Yale University Press, 2014), xxv.

34. See David Orwig, "Invasion of an Exotic Pest," in *Hemlock: A Forest Giant on the Edge*, ed. David R. Foster (New Haven, CT: Yale University Press, 2014), 120–35.

35. A copy of the brochure can be found at "Hemlock Hospice," Harvard Forest, 2017, https://harvardforest.fas.harvard.edu/sites/default/files/Hemlock%20Hospice%20Self-guided%20Map-PRINT.pdf.

36. See "Coronavirus in the U.S.: Latest Map and Case Count," *New York Times*, n.d.,

https://www.nytimes.com/.

37. This figure is current as of July 14, 2021, the date I last revised this sentence. For updated figures, see "COVID-19 Coronavirus Pandemic," Worldometer, November 3, 2020, https://www.worldometers.info/coronavirus/.
38. Peter Baker, "Amid a Rising Death Toll, Trump Leaves the Grieving to Others," *New York Times*, April 30, 2020, https://www.nytimes.com/; George Yancy and Judith Butler, "Judith Butler: Mourning Is a Political Act amid the Pandemic and Its Disparities," *Truthout* (blog), April 30, 2020, https://truthout.org/articles/judith-butler-mourning-is-a-political-act-amid-the-pandemic-and-its-disparities/.
39. "Those We've Lost," *New York Times*, https://www.nytimes.com/interactive/2020/obituaries/people-died-coronavirus-obituaries.html.
40. Michael Brice-Saddler, "Memorial Held outside White House in Solemn Reminder of More Than 200,000 Americans Killed by COVID-19," *Washington Post*, October 4, 2020, https://www.washingtonpost.com/.
41. On the relationship between mourning and the ethics of resistance, see R. Clifton Spargo, *The Ethics of Mourning: Grief and Responsibility in Elegiac Literature* (Baltimore: Johns Hopkins University Press, 2004), 1–38.
42. Jeff Orlowski, *Chasing Ice* (New York: Submarine Deluxe, 2012), documentary, 75 min.
43. Douglas Burton-Christie, "The Gift of Tears: Loss, Mourning, and the Work of Ecological Restoration," *Worldviews* 15, no. 1 (2011): 29–46.
44. Following Barbara Biesecker, I understand the task of critical rhetoric as one of "trac[ing] new lines of making sense by taking hold of the sign whose reference had been destabilized by and through those practices of resistance, lines that cut diagonally across and, thus, disrupt the social weave." See Barbara Biesecker, "Michel Foucault and the Question of Rhetoric," *Philosophy & Rhetoric* 25, no. 4 (1992): 361.

Works Cited

"About Us." n.d. Extinction Rebellion. Accessed November 1, 2020. https://rebellion.global/about-us/.

Abram, David. 1996. *The Spell of the Sensuous: Perception and Language in a More-Than-Human World*. New York: Vintage Books.

Ackroyd, Heather, and Dan Harvey. 2016. "Seeing Red.. Overdrawn." Ackroyd & Harvey (website). https://www.ackroydandharvey.com/seeing-red-overdrawn/.

Airhart, Ellen. 2021. "Taking Action Can Cure Your Climate Grief." *Scientific American*, March 23, 2021. https://www.scientificamerican.com/.

Albrecht, Glenn A. 2019. *Earth Emotions: New Words for a New World*. Ithaca, NY: Cornell University Press.

"Aldo Leopold Archives, The." n.d. University of Wisconsin–Madison Libraries. https://uwdc.library.wisc.edu/collections/aldoleopold/.

Andrews, Elaine. 1996. "The Last Word." *Nature* 381: 272.

Arendt, Hannah. 1977. *The Life of the Mind*. San Diego: Harcourt.

Aristotle. 1984. "Rhetoric." In *The Complete Works of Aristotle*, edited by Jonathan Barnes, translated by W. Rhys Roberts, II: 2152–2269. Princeton, NJ: Princeton University

Press.

Asafu-Adjaye, John, Linus Blomqvist, Stewart Brand, Barry Brook, and Ruth Defries. 2015. "An Ecomodernist Manifesto." https://static1.squarespace.com/static/5515d9f9e4b04d5c3198b7bb/t/552d37bbe4b07a7dd69fcdbb/1429026747046/An+Ecomodernist+Manifesto.pdf.

Atkinson, Jennifer. 2018. "Addressing Climate Grief Makes You a Badass, Not a Snowflake." *High County News*, May 29, 2018. https://www.hcn.org/articles/opinion-addressing-climate-grief-makes-you-a-badass-not-a-snowflake.

Attig, Thomas. 2011. *How We Grieve: Relearning the World*. 2nd ed. Oxford: Oxford University Press.

Audubon, John James. 1831. *Ornithological Biography, or An Account of the Habits of the Birds of the United States of America*. Philadelphia.

Azzarello, Nina. 2014. "Olafur Eliasson Moves 100 Tonnes of Ice to Copenhagen to Visualize Climate Change." *Designboom* (blog), October 24, 2014. https://www.designboom.com/art/olafur-eliasson-ice-watch-project-climate-change-greenland-copenhagen-10-24-2014/.

Baig, Fakiha. 2019. "Being Sad for the Planet: Edmontonians Share Experiences with Ecological Grief." *CBC News*, July 23, 2019. https://www.cbc.ca/news/.

Baker, Peter. 2020. "Amid a Rising Death Toll, Trump Leaves the Grieving to Others." *New York Times*, April 30, 2020. https://www.nytimes.com/.

Barnett, Joshua Trey. 2015. "Toxic Portraits: Resisting Multiple Invisibilities in the Environmental Justice Movement." *Quarterly Journal of Speech* 101, no. 2: 405–25.

Barnett, Joshua Trey. 2019a. "Irrational Hope, Phenological Writing, and the Prospects of Earthly Coexistence." *Communication and Critical/Cultural Studies* 16, no. 4: 382–91.

Barnett, Joshua Trey. 2019b. "Naming, Mourning, and the Work of Earthly Coexistence." *Environmental Communication* 13, no. 3: 287–99.

Barnett, Joshua Trey. 2020. "Grievable Water: Mourning the Animas River." In *Water, Rhetoric, and Social Justice: A Critical Confluence*, edited by Casey Schmitt, Christopher S. Thomas, and Theresa Castor, 273–91. Lanham, MD: Lexington.

Barnett, Joshua Trey. 2021a. "Ecological Moods: Feeling Our Way into Thinking Like Aldo Leopold." *ISLE* 28, no. 3: 1155–81.

Barnett, Joshua Trey. 2021b. "From Ecological Grief to *Gelassenheit*: Rhetorics of Sentiment and Science in *Hemlock*." *Environmental Communication* 15, no. 6: 787–97.

Barnett, Joshua Trey. 2021c. "Rhetoric for Earthly Coexistence: Imagining an Ecocentric Rhetoric." *Rhetoric & Public Affairs* 24, nos. 1–2: 365–78.

Barnett, Joshua Trey. 2021d. "Vigilant Mourning and the Future of Earthly Coexistence." In *Communicating in the Anthropocene: Intimate Relations*, edited by Alexa Dare and C. Vail Fletcher, 13–33. Lanham, MD: Lexington.

Barnett, Joshua Trey. n.d. "Eulogy Essay and Reading." UC-CSU NXTerra: Transformative Education for Climate Action (website). https://www.nxterra.orfaleacenter.ucsb.edu/wp-content/uploads/2020/09/Assignment-Climate-Change-Emotions-Eulogy-Essay-and-Presentation-Joshua-Barnett.pdf.

Barnett, Joshua Trey, and David Charles Gore. 2020. "Dwelling in the Anthropocene: Notes from Lake Superior." *Ethics & the Environment* 25, no. 1: 19–49.

Barr, Jessica Marion. 2017. "Auguries of Elegy: The Art and Ethics of Ecological Grieving." In *Mourning Nature: Hope at the Heart of Ecological Loss and Grief*, edited by Ashlee Cunsolo and Karen Landman, 190–226. Montreal: McGill–Queen's University Press.

Barthes, Roland. 2010. *Mourning Diary*. Translated by Richard Howard. New York: Hill and Wang.

Basso, Keith H. 1996. *Wisdom Sits in Places: Landscape and Language among the Western Apache*. Albuquerque: University of New Mexico Press.

Bellacasa, María Puig de la. 2017. *Matters of Care: Speculative Ethics in More Than Human Worlds*. Minneapolis: University of Minnesota Press.

Benjamin, Walter. 2008. "The Work of Art in the Age of Its Technological Reproducibility." In *The Work of Art in the Age of Its Technological Reproducibility, and Other Writings on Media*, edited by Michael W. Jennings, Brigid Doherty, and Thomas Y. Levin, translated by Edmund Jephcott, Rodney Livingstone, and Howard Eiland, 19–55. Cambridge, MA: Harvard University Press.

Biesecker, Barbara. 1992. "Michel Foucault and the Question of Rhetoric." *Philosophy & Rhetoric* 25, no. 4: 351–64.

Biesecker, Barbara. 2006. "Of Historicity, Rhetoric: The Archive as Scene of Invention." *Rhetoric & Public Affairs* 9, no. 1: 124–31.

Biesecker, Barbara. 2007. "No Time for Mourning: The Rhetorical Production of the Melancholic Citizen-Subject in the War on Terror." *Philosophy & Rhetoric* 40, no. 1: 147–69.

Birmingham, Robert A., and Amy L. Rosebrough. 2017. *Indian Mounds of Wisconsin*. 2nd ed. Madison: University of Wisconsin Press.

Bitzer, Lloyd F., and Edwin Black, eds. 1971. *The Prospect of Rhetoric: Report of the National Development Project*. Englewood Cliff, NJ: Prentice-Hall.

Black, Edwin. 1965. *Rhetorical Criticism: A Study in Method*. Madison: University of Wisconsin Press.

Bladow, Kyle, and Jennifer Ladino, eds. 2018. *Affective Ecocriticism: Emotion, Embodiment, Environment*. Lincoln: University of Nebraska Press.

Blair, Carole, Greg Dickinson, and Brian L. Ott. 2010. "Rhetoric/Memory/Place." In *Places of Public Memory: The Rhetoric of Museums and Memorials*, edited by Greg Dickinson, Carole Blair, and Brian L. Ott, 1–54. Tuscaloosa: University of Alabama Press.

Bradley, Nina L., A. Carl Leopold, John Ross, and Wellington Huffaker. 1999. "Phenological Changes Reflect Climate Change in Wisconsin." *Proceedings of the National Academy of Science of the United States of America* 96, no. 17: 9701–4.

Brault, Pascale-Anne, and Michael Naas. 2001. "To Reckon with the Dead: Jacques Derrida's Politics of Mourning." In *The Work of Mourning*, by Jacques Derrida, 1–30. Chicago: University of Chicago Press.

Braun, Sebastian F. 2017. "Mourning Ourselves and/as Our Relatives: Environment as Kinship." In *Mourning Nature: Hope at the Heart of Ecological Loss and Grief*, edited by Ashlee Cunsolo and Karen Landman, 64–91. Montreal: McGill–Queen's University Press.

Brice-Saddler, Michael. 2020. "Memorial Held outside White House in Solemn Reminder of More Than 200,000 Americans Killed by COVID-19." *Washington Post*, October 4, 2020. https://www.washingtonpost.com/.

Brunner, Elizabeth. 2019. *Environmental Activism, Social Media, and Protest in China: Becoming Activists over Wild Public Screens*. Lanham, MD: Lexington.

Burton-Christie, Douglas. 2011. "The Gift of Tears: Loss, Mourning, and the Work of Ecological Restoration." *Worldviews* 15, no. 1: 29–46.

Busby, Mattha. 2019. "Extinction Rebellion Activists Stage Die-In Protests across Globe." *Guardian*, April 27, 2019. https://www.theguardian.com/.

Busch, Akiko. 2013. *The Incidental Steward: Reflections on Citizen Science*. New Haven, CT: Yale University Press.

Busch, Akiko. 2019. *How to Disappear: Notes on Invisibility in a Time of Transparency*. New York: Penguin.

Butler, Judith. 1997. *The Psychic Life of Power: Theories in Subjection*. Stanford, CA: Stanford University Press.

Butler, Judith. 2000. *Antigone's Claim: Kinship between Life and Death*. New York: Columbia University Press.

Butler, Judith. 2004. *Precarious Life: The Powers of Mourning and Violence*. London: Verso.

Butler, Judith. 2009. *Frames of War: When Is Life Grievable*? London: Verso.

Butler, Judith. 2015. *Notes toward a Performative Theory of Assembly*. Cambridge, MA: Harvard University Press.

Care Collective. 2020. *The Care Manifesto: The Politics of Interdependence*. London: Verso.

Carson, Rachel. 1962. *Silent Spring*. Boston: Houghton Mifflin.

Case, Kristen. 2014. "Knowing as Neighboring: Approaching Thoreau's Kalendar." *J19* 2, no. 1: 107–29.

Case, Kristen. n.d. "About the Kalendar." Thoreau's Kalendar (website). https://thoreauskalendar.org/about.html.

Caswell, Michelle. 2016. "'The Archive' Is Not an Archives: Acknowledging the Intellectual Contributions of Archival Studies." *Reconstruction: Studies in Contemporary Culture* 16, no. 1. https://escholarship.org/uc/item/7bn4v1fk.

Cavanagh, Michaela. 2019. "It's Time to Talk about Ecological Grief." *Undark*, January 10, 2019. https://undark.org/2019/01/10/its-time-to-talk-about-ecological-grief/.

Charland, Maurice. 1987. "Constitutive Rhetoric: The Case of the *Peuple Québécois*." *Quarterly Journal of Speech* 73, no. 2: 133–50.

Chwałczyk, Franciszek. 2020. "Around the Anthropocene in Eighty Names: Considering the Urbanocene Proposition." *Sustainability* 12, no. 11: 1–33.

Clary-Lemon, Jennifer. 2019. *Planting the Anthropocene: Rhetorics of Natureculture*. Logan: Utah State University Press.

Clary-Lemon, Jennifer. 2020. "Examining Material Rhetorics of Species at Risk: Infrastructural Mitigations as Non-Human Arguments." *Enculturation: A Journal of Rhetoric, Writing, and Culture* 32. http://enculturation.net/material_rhetorics_species_at_risk.

Clayton, Susan, Christie Manning, and Caroline Hodge. 2014. "Beyond Storms and

Droughts: The Psychological Impacts of Climate Change." Washington, DC: American Psychological Association and ecoAmerica.

Clewell, Tammy. 2004. "Mourning beyond Melancholia: Freud's Psychoanalysis of Loss." *Journal of the American Psychoanalytic Association* 52, no. 1: 43–67.

Coates, Peter. 2007. *Strangers on the Land: American Perceptions of Immigrant and Invasive Species*. Berkeley: University of California Press.

Cockburn, Paige. 2020. "Climate Grief Expected to Be Widespread Soon but It's Still Not Openly Acknowledged." *ABC News*, August 18, 2020. https://www.abc.net.au/news/science/2020-08-19/the-danger-of-climate-grief-lies-in-the-fact-its-often-ignored/12563930.

Cook, Terry, and Joan M. Schwartz. 2002. "Archives, Records, and Power: From (Postmodern) Theory to (Archival) Performance." *Archival Science* 2, no. 3: 171–85.

Core Writing Team, Rajendra K. Pachauri, and Leo A. Meyer. 2014. "Climate Change 2014: Synthesis Report." Geneva: Intergovernmental Panel on Climate Change. https://www.ipcc.ch/site/assets/uploads/2018/02/SYR_AR5_FINAL_full.pdf.

Corrigan, Lisa M. 2020. *Black Feelings: Race and Affect in the Long Sixties*. Jackson: University Press of Mississippi.

"COVID-19 Coronavirus Pandemic." 2020. Worldometer (website). November 3, 2020. https://www.worldometers.info/coronavirus/.

Cox, Robert. 2007. "Nature's 'Crisis Disciplines': Does Environmental Communication Have an Ethical Duty?" *Environmental Communication* 1, no. 1: 5–20.

Craig, David. 1996. "The Last Word." *Nature* 381: 272.

Craps, Stef. 2020. "Introduction: Ecological Grief." *American Imago* 77, no. 1: 1–7.

Creel, Liz. 2003. "Ripple Effects: Population and Coastal Regions." Washington, DC: Population Review Bureau. https://www.prb.org/rippleeffectspopulationandcoastalregions/.

Crimmins, Theresa. 2020. "USA-NPN on The Weather Channel January 23, 2020." Interviewed by Jim Cantore and Stephanie Abrams. Aired January 23. YouTube video, 3:09. https://www.youtube.com/watch?v=NMzLSvdppjQ.

Crimp, Douglas. 1989. "Mourning and Militancy." *October* 51:3–18.

Crist, Eileen. 2013. "On the Poverty of Our Nomenclature." *Environmental Humanities* 3, no. 1: 129–47.

Cronon, William. 1995. "The Trouble with Wilderness: Or, Getting Back to the Wrong Nature." In *Uncommon Ground: Rethinking the Human Place in Nature*, edited by William Cronon, 69–90. New York: W. W. Norton.

Cronon, William. 1998. "'Only Connect . . .': The Goals of a Liberal Education." *American Scholar* 67, no. 4: 73–80.

Cross, John A. 2017. "Native American Landscapes in the Eastern United States." In *Ethnic Landscapes of America*, 21–46. Cham, Switzerland: Springer Nature.

Crutzen, Paul, and Eugene Stoermer. 2000. "The 'Anthropocene.'" *Global Change Newsletter* 41: 17–18.

Cunsolo, Ashlee, David Borish, Sherilee L. Harper, Jamie Snook, Inez Shiwak, Michelle Wood, and The Herd Caribou Steering Committee. 2020. "'You Can Never Replace the Caribou': Inuit Experiences of Ecological Grief from Caribou Declines." *American Imago* 77, no. 1: 31–59.

Cunsolo, Ashlee, and Neville R. Ellis. 2018. "Ecological Grief as a Mental Health Response to Climate Change-Related Loss." *Nature Climate Change* 8: 275–81.

Cunsolo, Ashlee, Sherilee L. Harper, Kelton Minor, Katie Hayes, Kimberly G. Williams, and Courtney Howard. 2020. "Ecological Grief and Anxiety: The Start of a Healthy Response to Climate Change?" *The Lancet* 4, no. 7: e261–63.

Cunsolo, Ashlee, and Karen Landman, eds. 2017a. *Mourning Nature: Hope at the Heart of Ecological Loss and Grief.* Montreal: McGill–Queen's University Press.

Cunsolo, Ashlee, and Karen Landman. 2017b. "To Mourn beyond the Human." In *Mourning Nature: Hope at the Heart of Ecological Loss and Grief*, edited by Ashlee Cunsolo and Karen Landman, 3–26. Montreal: McGill–Queen's University Press.

Cunsolo Willox, Ashlee. 2014. *Attutauniujuk Nunami/Lament for the Land.* http://www.lamentfortheland.ca/film/.

Curtin, Deane. 1991. "Toward an Ecological Ethic of Care." *Hypatia* 6, no. 1: 60–74.

Cvetkovich, Ann. 2003. *An Archive of Feelings: Trauma, Sexuality, and Lesbian Public Cultures.* Durham, NC: Duke University Press.

Cvetkovich, Ann. 2012. *Depression: A Public Feeling.* Durham, NC: Duke University Press.

DeLaure, Marilyn. 2020. "Performing Loss: Sonic Rhetoric in Maya Lin's *What Is Missing*?" *Liminalities* 16, no. 1: 1–32.

Deleuze, Gilles. 1994. *Difference and Repetition.* Translated by Paul Patton. New York:

Columbia University Press.

DeLuca, Kevin. 1999. "Articulation Theory: A Discursive Grounding for Rhetorical Practice." *Philosophy & Rhetoric* 32, no. 4: 334–48.

DeLuca, Kevin Michael. 1999a. *Image Politics: The New Rhetoric of Environmental Activism*. New York: Guilford.

DeLuca, Kevin Michael. 1999b. "Unruly Arguments: The Body Rhetoric of Earth First!, ACT UP, and Queer Nation." *Argumentation and Advocacy* 36, no. 1: 9–21.

DeLuca, Kevin Michael. 2006. "The Speed of Immanent Images: The Dangers of Reading Photographs." In *Visual Communication: Perception, Rhetoric, and Technology*, edited by Diane S. Hope, 79–90. Cresskill, NJ: Hampton.

Demarée, Gaston R., and This Rutishauser. 2011. "From 'Periodical Observations' to 'Anthochronology' and 'Phenology': The Scientific Debate between Adolphe Quetelet and Charles Morren on the Origin of the Word 'Phenology.'" *International Journal of Biometeorology* 55, no. 6: 753–61.

DeMello, Margo. 2016a. "Introduction." In *Mourning Animals: Rituals and Practices Surrounding Animal Death*, edited by Margo DeMello, xvii–xxv. East Lansing: Michigan State University Press.

DeMello, Margo, ed. 2016b. *Mourning Animals: Rituals and Practices Surrounding Animal Death*. East Lansing: Michigan State University Press.

Derrida, Jacques. 1981. *Dissemination*. Translated by Barbara Johnson. Chicago: University of Chicago Press.

Derrida, Jacques. 1982a. "Différance." In *Margins of Philosophy*, translated by Alan Bass, 1–27. Chicago: University of Chicago Press.

Derrida, Jacques. 1982b. "Signature Event Context." In *Margins of Philosophy*, translated by Alan Bass, 307–30. Chicago: University of Chicago Press.

Derrida, Jacques. 1995. *Archive Fever: A Freudian Impression*. Translated by Eric Prenowitz. Chicago: University of Chicago Press.

Derrida, Jacques. 1997. *The Politics of Friendship*. Translated by George Collins. London: Verso.

Derrida, Jacques. 2001a. "By Force of Mourning." In *The Work of Mourning*, translated by Pascale-Anne Brault and Michael Naas. Chicago: University of Chicago Press.

Derrida, Jacques. 2001b. "The Taste of Tears." In *The Work of Mourning*, edited by

Pascale-Anne Brault and Michael Naas, 107–10. Chicago: University of Chicago Press.

Derrida, Jacques. 2008. *The Animal That Therefore I Am*. New York: Fordham University Press.

Derrida, Jacques, and Elizabeth Roudinesco. 2004. "Violence against Animals." In *For What Tomorrow . . . A Dialogue*, translated by Jeff Fort, 62–76. Stanford, CA: Stanford University Press.

"Designer of the Vietnam Veterans Memorial: Maya Lin." 2020. Vietnam Veterans Memorial Fund. https://www.vvmf.org/About-The-Wall/history-of-the-vietnam-veterans-memorial/Maya-Lin/.

Dimick, Sarah. 2018. "Disordered Environmental Time: Phenology, Climate Change, and Seasonal Form in the Work of Henry David Thoreau and Aldo Leopold." *ISLE* 25, no. 4: 700–721.

Dohms-Harter, Elizabeth. 2020. "'A Sculpted Land': Wisconsin's Effigy Mounds Connected People to Spirits, Humanity." *Wisconsin Public Radio* (blog). September 2, 2020. https://www.wpr.org/sculpted-land-wisconsins-effigy-mounds-connected-people-spirits-humanity.

Doka, Kenneth J., ed. 1989. *Disenfranchised Grief: Recognizing Hidden Sorrow*. Lexington, MA: Lexington.

Dunbar-Ortiz, Roxanne. 2014. *An Indigenous Peoples' History of the United States*. Boston: Beacon.

Dunbar-Ortiz, Roxanne, and Dina Gilio-Whitaker. 2016. *"All the Real Indians Died Off" and 20 Other Myths About Native Americans*. Boston: Beacon.

Dunn, Thomas R. 2011. "Remembering 'A Great Fag': Visualizing Public Memory and the Construction of Queer Space." *Quarterly Journal of Speech* 97, no. 4: 435–60.

Edbauer, Jenny. 2005. "Unframing Models of Public Distribution: From Rhetorical Situation to Rhetorical Ecologies." *Rhetoric Society Quarterly* 35, no. 4: 5–24.

Ellwood, Elizabeth R., Stanley A. Temple, Richard B. Primack, Nina L. Bradley, and Charles C. Davis. 2013. "Record-Breaking Early Flowering in the Eastern United States." *PLOS One* 8, no. 1: 1–9.

Endres, Danielle. 2020. "Environmental Criticism." *Western Journal of Communication* 84, no. 3: 314–31.

Engels, Jeremy David. 2015. *The Politics of Resentment: A Genealogy*. University Park:

Pennsylvania State University Press.

Engels, Jeremy David. 2018. *The Art of Gratitude*. Albany: State University of New York Press.

Estes, Nick. 2019. *Our History Is the Future: Standing Rock versus the Dakota Access Pipeline, and the Long Tradition of Indigenous Resistance*. London: Verso.

Evernden, Neil. 1992. *The Social Creation of Nature*. Baltimore: Johns Hopkins University Press.

Fahys, Judy. 2017. "First Step to 'Eco-Grieving' over Climate Change? Admit There's a Problem." National Public Radio, April 22, 2017. https://www.npr.org/.

Farrell, Thomas B. 2008. "The Weight of Rhetoric: Studies in Cultural Delirium." *Philosophy & Rhetoric* 41, no. 4: 467–87.

Fatsis, Stefan. 2015. "Panic at the Dictionary." *New Yorker*, January 30, 2015. https://www.newyorker.com/.

"15 Million Records Submitted to the National Phenology Database." 2018. USA National Phenology Network. December 17, 2018. https://www.usanpn.org/node/35012.

Finnegan, Cara. 2006. "Communication as Vision." In *Communication As . . . Perspectives on Theory*, edited by Gregory J. Shepherd, Jeffrey St. John, and Ted Striphas, 60–66. Thousand Oaks, CA: SAGE.

Flood, Alison. 2015. "Oxford Junior Dictionary's Replacement of 'Natural' Words With 21st-Century Terms Sparks Outcry." *Guardian*, January 13, 2015. https://www.theguardian.com/.

Floyd, Ted. 2008. *Field Guide to the Birds of North America*. New York: Harper Collins.

Foster, David R., ed. 2014a. *Hemlock: A Forest Giant on the Edge*. New Haven, CT: Yale University Press.

Foster, David R. 2014b. "Hemlock's Future in the Context of Its Past." In *Hemlock: A Forest Giant on the Edge*, edited by David R. Foster, 1–10. New Haven, CT: Yale University Press.

Foster, David R. 2014c. "Preface." In *Hemlock: A Forest Giant on the Edge*, edited by David R. Foster, xxi–xxviii. New Haven, CT: Yale University Press.

Foucault, Michel. 1995. "The Body of the Condemned." In *Discipline and Punish: The Birth of the Prison*, translated by Alan Sheridan, 3–31. New York: Vintage.

Freedman, Eric. 2011. "Extinction Is Forever: A Quest for the Last Known Survivors." *Earth Island Journal*, Autumn 2011. https://www.earthisland.org/journal/index.php/

magazine/entry/extinction_is_forever/.

Freud, Sigmund. 1953. "Mourning and Melancholia." In *Collected Papers: Volume IV*, edited by Joan Riviere, 152–70. London: Hogarth.

Freud, Sigmund. 1960. *The Ego and the Id: Standard Edition*. Edited by James Strachey. Translated by Joan Riviere. New York: W. W. Norton.

Gilio-Whitaker, Dina. 2019. *As Long as Grass Grows: The Indigenous Fight for Environmental Justice; From Colonization to Standing Rock*. Boston: Beacon.

Goleman, Daniel. 1995. *Emotional Intelligence: Why It Can Matter More Than IQ*. New York: Bantam.

Gordon, Timothy A. C., Andrew N. Radford, and Stephen D. Simpson. 2019. "Grieving Environmental Scientists Need Support." *Science* 366, no. 6462: 193.

Gore, David Charles. 2019. *The Voice of the People: Political Rhetoric in the Book of Mormon*. Provo, UT: Brigham Young University.

Graves, Robert, and Didier Madoc-Jones. n.d. "Images." Postcards from the Future (website). http://www.postcardsfromthefuture.com/postcard_images/.

Graves, Robert, and Didier Madoc-Jones. n.d. "The Project." Postcards from the Future (website). http://www.postcardsfromthefuture.com/the-project/.

"Great Acceleration." 2015. International Geosphere-Biosphere Programme. http://www.igbp.net/globalchange/greatacceleration.4.1b8ae20512db692f2a680001630.html.

Greenberg, Joel. 2014. *A Feathered River across the Sky: The Passenger Pigeon's Flight to Extinction*. New York: Bloomsbury.

"Grief vs. Mourning: What's the Difference?" 2021. The Recovery Village (website). January 4, 2021. https://www.therecoveryvillage.com/mental-health/grief/related/grief-vs-mourning/.

Hamilton, Clive. 2017. *Defiant Earth: The Fate of Humans in the Anthropocene*. Cambridge, UK: Polity.

Haraway, Donna. 2008. *When Species Meet*. Minneapolis: University of Minnesota Press.

Haraway, Donna. 2016. *Staying with the Trouble: Making Kin in the Cthulucene*. Durham, NC: Duke University Press.

Harris, Darcy L., ed. 2020. *Non-Death Loss and Grief: Context and Clinical Implications*. New York: Routledge.

Harrison, Robert Pogue. 1993. *Forests: The Shadow of Civilization*. Chicago: University of

Chicago Press.

Harrison, Robert Pogue. 2003. *The Dominion of the Dead*. Chicago: University of Chicago Press.

Hart, Adam G., William S. Carpenter, Estelle Hlustik-Smith, Matt Reed, and Anne E. Goodenough. 2018. "Testing the Potential of Twitter Mining Methods for Data Acquisition: Evaluating Novel Opportunities for Ecological Research in Multiple Taxa." *Methods in Ecology and Evolution* 9, no. 11: 2194–2205.

Haskins, Ekaterina. 2007. "Between Archive and Participation: Public Memory in a Digital Age." *Rhetoric Society Quarterly* 37, no. 4: 401–22.

Hass, Robert. 2006. "Ford." In *Home Ground: A Guide to the American Landscape*, edited by Barry Lopez and Debra Gwartney, 186–87. San Antonio: Trinity University Press.

Hatfield, Samantha Chisholm, Elizabeth Marino, Kyle Powys Whyte, and Philip W. Mote. 2018. "Indian Time: Time, Seasonality, and Culture in Traditional Ecological Knowledge of Climate Change." *Ecological Processes* 7, no. 25: 1–11.

Head, Lesley, and Theresa Harada. 2017. "Keeping the Heart a Long Way from the Brain: The Emotional Labour of Climate Scientists." *Emotion, Space and Society* 24: 34–41.

Heidegger, Martin. 2010. *Being and Time*. Translated by Joan Stambaugh. Albany: State University of New York Press.

Heigl, Florian, Barbara Kieslinger, Katharina T. Paul, Julia Uhlik, and Daniel Dörler. 2019. "Toward an International Definition of Citizen Science." *Proceedings of the National Academy of Science of the United States of America* 116, no. 17: 8089–92.

Held, Virginia. 2006. *The Ethics of Care: Personal, Political, and Global*. Oxford: Oxford University Press.

"Hemlock Hospice." 2017. Harvard Forest. https://harvardforest.fas.harvard.edu/sites/default/files/Hemlock%20Hospice%20Self-guided%20Map-PRINT.pdf.

"Hemlock Hospice." n.d. Harvard Forest. https://harvardforest.fas.harvard.edu/hemlock-hospice.

Hill, Julia Butterfly. 2001. *The Legacy of Luna: The Story of a Tree, a Woman, and the Struggle to Save the Redwoods*. San Francisco: Harper Collins.

Hineline, Mark L. 2018. *Ground Truth: A Guide to Tracking Climate Change at Home*. Chicago: University of Chicago Press.

Hogan, Linda. 1995. *Dwellings: A Spiritual History of the Living World*. New York: W. W.

Norton.

Holson, Laura M. 2019. "Iceland Mourns Loss of a Glacier by Posting a Warning about Climate Change." *New York Times*, August 19, 2019. https://www.nytimes.com/.

Holthaus, Eric. 2018. "If You're Suffering from Climate Grief, You're Not Alone." *Grist* (blog). October 15, 2018. https://grist.org/article/climate-grief-un-ipcc-report/.

Hopp, Richard J. 1974. "Plant Phenology Observation Networks." In *Phenology and Seasonality Modeling*, edited by Helmut Lieth, 25–43. New York: Springer.

Houser, Heather. 2014. *Ecosickness in Contemporary U.S. Fiction: Environment and Affect*. New York: Columbia University Press.

Hu, Jane C. 2019. "How Can You Tell When a Glacier Is Dead?" *Slate*, July 24, 2019. https://slate.com/.

Huebener, Paul. 2020. *Nature's Broken Clocks: Reimagining Time in the Face of Environmental Crisis*. Regina, CA: University of Regina Press.

Ice Watch. 2015. Studio Olafur Eliasson.

Ice Watch, 2014. Olafur Eliasson. https://olafureliasson.net/archive/artwork/WEK109190/ice-watch.

Ingraham, Chris. 2020. *Gestures of Concern*. Durham, NC: Duke University Press.

"IUCN Red List of Threatened Species, The." 2020. IUCN Red List. https://www.iucnredlist.org.

Jamail, Dahr. 2019. *The End of Ice: Bearing Witness and Finding Meaning in the Path of Climate Change*. New York: The New Press.

Jefferson, Thomas. 2013. *The Papers of Thomas Jefferson, Volume 40: 4 March to 10 July 1803*. Edited by Barbara B. Oberg. Princeton, NJ: Princeton University Press.

Jensen, Tim. 2019. *Ecologies of Guilt in Environmental Rhetorics*. London: Palgrave Macmillan.

Johnson, Davi. 2007. "Martin Luther King Jr.'s 1963 Birmingham Campaign as Image Event." *Rhetoric & Public Affairs* 10, no. 1: 1–25.

Jonze, Tim. 2018. "Icebergs Ahead! Olafur Eliasson Brings the Frozen Fjord to Britain." *Guardian*, December 11, 2018. https://www.theguardian.com/.

Jørgensen, Dolly. 2017. "Endling, the Power of the Last in an Extinction-Prone World." *Environmental Philosophy* 14, no. 1: 119–38.

Jørgensen, Dolly. 2019. *Recovering Lost Species in the Modern Age: Histories of Longing and Belonging*. Cambridge, MA: MIT Press.

Jørgensen, Dolly. "Naming the Victims of the Sixth Mass Extinction." Penn Program in Environmental Humanities Lab. Posted April 15, 2020. YouTube video titled "Naming Extinction with Dolly Jørgensen," 19:25. https://www.youtube.com/watch?v=adWfrKYH_GM&feature=youtu.be.

Kassam, Karim-Aly S., Morgan L. Ruelle, Cyrus Samimi, Antonio Trabucco, and Jianchu Xu. 2018. "Anticipating Climatic Variability: The Potential of Ecological Calendars." *Human Ecology* 46, no. 2: 249–57.

Kennerly, Michele. 2010. "Getting Carried Away: How Rhetorical Transport Gets Judgment Going." *Rhetoric Society Quarterly* 40, no. 3: 269–91.

Kenote, Thomas Richmond. 2020. "Indigenous Phenology: An Interdisciplinary Case Study on Indigenous Phenological Knowledge on the Menominee Nation Forest." Master's Thesis, University of Minnesota.

Kershner, Bruce, Daniel Matthews, Gil Nelson, Richard Spellenberg, Terry Purinton, Andrew Block, Gerry Moore, and John W. Thieret. 2008. *Field Guide to Trees of North America*. New York: Sterling.

Kevorkian, Kriss. 2004. "Environmental Grief: Hope and Healing." PhD diss., Union Institute and University.

Kimmerer, Robin Wall. 2003. *Gathering Moss: A Natural and Cultural History of Mosses*. Corvalis, OR: Oregon State University Press.

Kimmerer, Robin Wall. 2013. *Braiding Sweetgrass: Indigenous Wisdom, Scientific Knowledge, and the Teachings of Plants*. Minneapolis: Milkweed.

Kirkby, Joan. 2006. "'Remembrance of the Future': Derrida on Mourning." *Social Semiotics* 16, no. 3: 461–72.

Kirkman, Robert. 2007. "A Little Knowledge of Dangerous Things: Human Vulnerability in a Changing Climate." In *Merleau-Ponty and Environmental Philosophy: Dwelling on the Landscape of Thought*, edited by Suzanne L. Cataldi and William S. Hamrick, 19–35. Albany: State University of New York Press.

Kolbert, Elizabeth. 2014. *The Sixth Extinction: An Unnatural History*. New York: Picador.

Kolbert, Elizabeth. 2021. *Under a White Sky: The Nature of the Future*. New York: Crown.

Kretz, Lisa. 2017. "Emotional Solidarity: Ecological Emotional Outlaws Mourning Environmental Loss and Empowering Positive Change." In *Mourning Nature: Hope at the Heart of Ecological Loss and Grief*, edited by Ashlee Cunsolo and Karen Landman,

258–91. Montreal: McGill–Queen's University Press.

Kübler-Ross, Elizabeth, and David Kessler. 2005. *On Grief and Grieving: Finding the Meaning of Grief through the Five Stages of Loss*. New York: Scribner.

Ladino, Jennifer K. 2019. *Memorials Matter: Emotion, Environment, and Public Memory at American Historical Sites*. Reno: University of Nevada Press.

LaDuke, Winona. 2006. "Three Affiliated Tribes at a Crossroads: Which Energy Path?" *Indian Country Today*, April 6, 2006. https://indiancountrytoday.com/archive/three-affiliated-tribes-at-a-crossroads-which-energy-path.

LaDuke, Winona. 2014. "Crow and Lummi, Dirty Coal and Clean Fishing." *Indian Country Today*, January 15, 2014. https://indiancountrytoday.com/archive/crow-lummi-dirty-coal-clean-fishing.

LaDuke, Winona. 2015. *All Our Relations: Native Struggles for Land and Life*. Chicago: Haymarket.

LaDuke, Winona. 2016. *Recovering the Sacred: The Power of Naming and Claiming*. Chicago: Haymarket.

Landau, Jamie, and Bethany Keeley-Jonker. 2018. "Conductor of Public Feelings: An Affective-Emotional Rhetorical Analysis of Obama's National Eulogy in Tucson." *Quarterly Journal of Speech* 104, no. 2: 166–88.

Larsen, Soren C., and Jay T. Johnson. 2017. *Being Together in Place: Indigenous Coexistence in a More Than Human World*. Minneapolis: University of Minnesota Press.

Laurens, Emily. 2014. "Remembrance Day for Lost Species." *Emily Laurens* (blog), November 28, 2014. https://emilylaurens.wordpress.com/2014/11/28/remembrance-day-for-lost-species/.

Lawson, Michael L. 1982. *Dammed Indians: The Pick-Sloan Act and the Missouri River Sioux, 1944–1980*. Norman: University of Oklahoma Press.

Lear, Jonathan. 2006. *Radical Hope: Ethics in the Face of Cultural Destruction*. Cambridge, MA: Harvard University Press.

"Learn How to Observe." n.d. USA National Phenology Network. https://www.usanpn.org/nn/guidelines.

Leopold, Aldo. 1947. "On a Monument to the Pigeon." In *Silent Wings: A Memorial to the Passenger Pigeon*, edited by Walter Edwin Scott, 3–5. Madison: Wisconsin Society for Ornithology.

Leopold, Aldo. 1949a. *A Sand County Almanac and Sketches Here and There*. New York: Oxford University Press.

Leopold, Aldo. 1949b. "Thinking Like a Mountain." In *A Sand County Almanac and Sketches Here and There*, 129–33. New York: Oxford University Press.

Leopold, Aldo. 1966. "The Round River." In *A Sand County Almanac with Essays on Conservation from Round River*, 188–202. New York: Ballantine.

Leopold, Aldo. 1986. *Game Management*. Reprint ed. Madison: University of Wisconsin Press.

Leopold, Aldo, and Sara Elizabeth Jones. 1947. "A Phenological Record for Sauk and Dane Counties, Wisconsin, 1935–1945." *Ecological Monographs* 17, no. 1: 81–112.

"Leopold Shack Journal, 1946–1948." n.d. Box 003, folder 003. Aldo Leopold Papers, University of Wisconsin–Madison Libraries. http://digital.library.wisc.edu/1711.dl/AldoLeopold.

Lertzman, Renee. 2015. *Environmental Melancholia: Psychoanalytic Dimensions of Engagement*. London: Routledge.

Linderman, Frank B. 2002. *Plenty-Coups: Chief of the Crows*. Lincoln, NE: Bison.

"London Futures / Metro and Museum of London Competition." 2010. Flickr. https://www.flickr.com/groups/londonfuturesmetro/.

Loveridge, Andrew. 2018. *Lion Hearted: The Life and Death of Cecil and the Future of Africa's Iconic Cats*. New York: Regan Arts.

Lucaites, John Louis, and Celeste Michelle Condit. 1999. "Introduction." In *Contemporary Rhetorical Theory: A Reader*, edited by John Louis Lucaites, Celeste Michelle Condit, and Sally Caudill, 1–18. New York: Guilford.

Macfarlane, Robert. 2015. *Landmarks*. New York: Penguin.

Macfarlane, Robert, and Jackie Morris. 2018. *The Lost Words: A Spell Book*. Toronto: House of Anansi.

Malatino, Hil. 2020. *Trans Care*. Minneapolis: University of Minnesota Press.

Malm, Andreas, and Alf Hornborg. n.d. "The Geology of Mankind? A Critique of the Anthropocene Narrative." *Anthropocene Review* 1, no. 1: 62–69.

Mark, Andrew. 2016. "Don't Organize, Mourn: Environmental Loss and Musicking." *Ethics & the Environment* 21, no. 2: 51–77.

Marlon, Jennifer, Peter Howe, Matto Mildenberger, Anthony Leiserowitz, and Xinran Wang. 2019. "Yale Climate Opinion Maps 2019." Yale Program on Climate Change

Communication, September 17, 2019. https://climatecommunication.yale.edu/visualizations-data/ycom-us/.

Marshall, George. 2010. "Fantasy Images of Climate Migration Will Fuel Existing Prejudices." *Guardian*, October 27, 2010. https://www.theguardian.com/.

Marshall, Nadine, William Neil Adger, Claudia Benham, Katrina Brown, Matthew I. Curnock, Georgina G. Gurney, Paul Marshall, Petina L. Pert, and Lauric Thiault. 2019. "Reef Grief: Investigating the Relationship between Place Meanings and Place Change on the Great Barrier Reef, Australia." *Sustainability Science* 14: 579–87.

McCaig, Amy. 2019. "Saying Goodbye to Okjökull: Howe Reflects on Glacier Memorial Ceremony." *Rice University News and Media Relations* (blog), September 10, 2019. https://news.rice.edu/2019/09/10/saying-goodbye-to-okjokull-howe-reflects-on-glacier-memorial-ceremony/.

McClure, Mark S., Scott M. Salom, and Kathleen S. Shields. 1996. "Hemlock Woolly Adelgid." FHTET-96-35. Morgantown, WV: USDA Forest Health Technology Enterprise Team.

McDougall, Dan. 2019. "'Ecological Grief': Greenland Residents Traumatised by Climate Emergency." *Guardian*, August 12, 2019. https://www.theguardian.com/.

McGee, Michael Calvin. 1990. "Text, Context, and the Fragmentation of Contemporary Culture." *Western Journal of Communication* 54, no. 3: 274–89.

McKerrow, Raymie. 1989. "Critical Rhetoric: Theory and Praxis." *Communication Monographs* 56, no. 2: 91–111.

McKibben, Bill. 2010. *Eaarth: Making a Life on a Tough New Planet*. New York: St. Martin's Press.

Meine, Curt. 2010. *Aldo Leopold: His Life and Work*. 2nd ed. Madison: University of Wisconsin Press.

Meloche, Katherine. 2018. "Mourning Landscapes and Homelands: Indigenous and Non-Indigenous Peoples' Ecological Griefs." *Canadian Mountain Network* (blog), August 30, 2018. https://canadianmountainnetwork.ca/mourning-landscapes-and-homelands-indigenous-and-non-indigenous-peoples-ecological-griefs/.

Menning, Nancy. 2017. "Environmental Mourning and the Religious Imagination." In *Mourning Nature: Hope at the Heart of Ecological Loss and Grief*, edited by Ashlee Cunsolo and Karen Landman, 39–63. Montreal: McGill–Queen's University Press.

"Mission." 2018. Harvard Forest. https://harvardforest.fas.harvard.edu/mission.

Moore, Jason W. 2017. "The Capitalocene, Part I: On the Nature and Origins of Our Ecological Crisis." *The Journal of Peasant Studies* 44, no. 3: 594–630.

Morris, Charles E., III. 2006. "Archival Queer." *Rhetoric & Public Affairs* 9, no. 1: 145–51.

Mortimer-Sandilands, Catriona. 2010. "Melancholy Natures, Queer Ecologies." In *Queer Ecologies: Sex, Nature, Politics, Desire*, edited by Catriona Mortimer-Sandilands and Bruce Erickson, 331–58. Bloomington: Indiana University Press.

Morton, Timothy. 2010a. "The Dark Ecology of Elegy." In *The Oxford Handbook of the Elegy*, edited by Karen Weisman, 251–71. Oxford: Oxford University Press.

Morton, Timothy. 2010b. *The Ecological Thought*. Cambridge, MA: Harvard University Press.

Morton, Timothy. 2018. *Being Ecological*. Cambridge, MA: MIT Press.

Mossner, Alexa Weik von. 2017. *Affective Ecologies: Empathy, Emotion, and Environmental Narrative*. Columbus: Ohio State University Press.

Muller, S. Marek. 2020. *Impersonating Animals: Rhetoric, Ecofeminism, and Animal Rights Law*. East Lansing: Michigan State University Press.

New York Times. n.d. "Covid in the U.S.: Latest Map and Case Count." https://www.nytimes.com/interactive/2020/us/coronavirus-us-cases.html.

New York Times. n.d. "Those We've Lost." Accessed November 3, 2020. https://www.nytimes.com/interactive/2020/obituaries/people-died-coronavirus-obituaries.html.

Nietzsche, Friedrich. 1999. "On Truth and Lying in a Non-Moral Sense." In *The Birth of Tragedy and Other Writings*, edited by Raymond Guess and Ronald Speirs, 178–92. Cambridge: Cambridge University Press.

Nijhuis, Michelle. 2017. "What Do You Call the Last of a Species?" *New Yorker*, March 2, 2017. https://www.newyorker.com/.

Norgaard, Kari Marie. 2011. *Living in Denial: Climate Change, Emotions, and Everyday Life*. Cambridge, MA: MIT Press.

Nowacki, Gregory J., and Marc D. Abrams. 1994. "Forest Composition, Structure, and Disturbance History of the Alan Seeger Natural Area, Huntington County, Pennsylvania." *Bulletin of the Torrey Botanical Club* 121, no. 3: 277–91.

Ohl, Michael. 2018. *The Art of Naming*. Translated by Elisabeth Lauffer. Cambridge, MA: MIT Press.

Olson, Lester C., Cara A. Finnegan, and Diane S. Hope. 2008. "Visual Rhetoric in

Communication: Continuing Questions and Contemporary Issues." In *Visual Rhetoric: A Reader in Communication and American Culture*, edited by Lester C. Olson, Cara A. Finnegan, and Diane S. Hope, 1–14. Los Angeles: SAGE Publishing.

Orlowski, Jeff, dir. 2012. *Chasing Ice*. New York: Submarine Deluxe. Documentary.

Orwig, David. 2014a. "An Iconic Species." In *Hemlock: A Forest Giant on the Edge*, edited by David R. Foster, 11–24. New Haven, CT: Yale University Press.

Orwig, David. 2014b. "Invasion of an Exotic Pest." In *Hemlock: A Forest Giant on the Edge*, edited by David R. Foster, 120–35. New Haven, CT: Yale University Press.

Paradis, Annie, Joe Elkinton, Katharine Hayhoe, and John Buonaccorsi. 2008. "Role of Winter Temperature and Climate Change on the Survival and Future Range Expansion of the Hemlock Woolly Adelgid (*Adelges tsugae*) in Eastern North America." *Mitigation and Adaptation Strategies for Global Change* 13, no. 5–6: 541–54.

"Passenger Pigeon, The." 2001. Smithsonian Institution. https://www.si.edu/spotlight/passenger-pigeon.

Peeples, Jennifer. 2011. "Toxic Sublime: Imaging Contaminated Landscapes." *Environmental Communication* 5, no. 4: 373–92.

"Peer-Reviewed Results." n.d. USA National Phenology Network. https://www.usanpn.org/pubs/results.

Peters, John Durham. 2015. *The Marvelous Clouds: Toward a Philosophy of Elemental Media*. Chicago: University of Chicago Press.

Petrauski, Lori, Sheldon F. Owen, George D. Constantz, and James T. Anderson. 2019. "Changes in Flowering Phenology of *Cardamine concatenata* and *Erythronium americanum* over 111 Years in the Central Appalachians." *Plant Ecology* 220, no. 9: 817–28.

Pezzullo, Phaedra C. 2007. *Toxic Tourism: Rhetorics of Pollution, Travel, and Environmental Justice*. Tuscaloosa: University of Alabama Press.

Pezzullo, Phaedra C. 2016. "Unearthing the Marvelous: Environmental Imprints in Rhetorical Criticism." *Review of Communication* 16, no. 1: 25–42.

Pezzullo, Phaedra C. 2017. "Environment." In *Oxford Research Encyclopedia of Communication*, edited by Jon F. Nussbaum, 1–20. Oxford: Oxford University Press.

Pezzullo, Phaedra C., and Robert Cox. 2018. *Environmental Communication and the Public Sphere*. 5th ed. Los Angeles: SAGE Publishing.

Plato. 2004. *Gorgias*. Translated by Walter Hamilton and Chris Emlyn-Jones. Revised ed. New York: Penguin Books.

Plato. 2005. *Phaedrus*. Translated by Christopher Rowe. New York: Penguin.

Porter, Eliot. 1963. *The Place No One Knew: Glen Canyon on the Colorado*. San Francisco: Sierra Club.

Primack, Richard B. 2014. *Walden Warming: Climate Change Comes to Thoreau's Woods*. Chicago: University of Chicago Press.

Rae, Patricia. 2003. "Double Sorrow: Proleptic Elegy and the End of Arcadianism in 1930s Britain." *Twentieth Century Literature* 49, no. 2: 246–75.

Randall, Rosemary. 2009. "Loss and Climate Change: The Cost of Parallel Narratives." *Ecopsychology* 1, no. 3: 118–29.

Ray, Sarah Jaquette. 2013. *The Ecological Other: Environmental Exclusion in American Culture*. Tucson: University of Arizona Press.

Ray, Sarah Jaquette. 2018. "Coming of Age at the End of the World: The Affective Arc of Undergraduate Environmental Studies Curricula." In *Affective Ecocriticism: Emotion, Embodiment, Environment*, edited by Kyle Bladow and Jennifer Ladino, 299–319. Lincoln: University of Nebraska Press.

Ray, Sarah Jaquette. 2020. *A Field Guide to Climate Anxiety: How to Keep Your Cool on a Warming Planet*. Oakland: University of California Press.

Redmalm, David. 2015. "Pet Grief: When Is Non-Human Life Grievable?" *Sociological Review* 63, no. 1: 19–35.

Rose, Deborah Bird. 2011. *Wild Dog Dreaming: Love and Extinction*. Charlottesville: University of Virginia Press.

Rosenfeld, Jordan. 2016. "Facing Down 'Environmental Grief.'" *Scientific American*, July 21, 2016. https://www.scientificamerican.com/.

Savoy, Lauret. 2015. *Trace: Memory, History, Race, and the American Landscape*. Berkeley, CA: Counterpoint.

Schmitt, Casey R. 2015. "Contours of the Land: Place-as-Rhetoric and Native American Effigy Mounds." *Western Journal of Communication* 79, no. 3: 307–26.

Schneider, Haley. 2021. "Public Futures: The Rhetoric of Sustainability and Survival in Philadelphia's Environmental Initiatives." Lecture presented at the Center for Democratic Deliberation Lecture Series, Pennsylvania State University, University Park,

PA, March 24.

"School History." 2017. Haskell Indian Nations University. https://www.haskell.edu/about/history/.

Scott, Gertrude M. 1947. "Society Meets in May at Madison." *The Passenger Pigeon* 9, no. 2: 54–62.

Scott, Walter Edwin, ed. 1947. *Silent Wings: A Memorial to the Passenger Pigeon*. Madison: Wisconsin Society for Ornithology.

Scranton, Roy. 2015. *Learning to Die in the Anthropocene*. San Francisco: City Lights Books.

"Shack, The." 2020. Aldo Leopold Foundation. https://www.aldoleopold.org/visit/the-shack/.

Shanahan, Mike. 2013. "Unhappy Endlings: What Tales of the Last Days of Extinction and Dying Species Can Bring to Our Own Story." *Under the Banyan* (blog), April 2, 2013. https://underthebanyan.blog/2013/04/02/unhappy-endlings-what-tales-of-the-last-days-of-extinct-and-dying-species-can-bring-to-our-own-story/.

"*Silphium laciniatum* L., Compass Plant." n.d. USDA Natural Resources Conservation Service. https://plants.usda.gov/core/profile?symbol=SILA3.

Smith, Mark. 1996. "The Last Word." *Nature* 381:272.

Spargo, R. Clifton. 2004. *The Ethics of Mourning: Grief and Responsibility in Elegiac Literature*. Baltimore: Johns Hopkins University Press.

Stanescu, James. 2012. "Species Trouble: Judith Butler, Mourning, and the Precarious Lives of Animals." *Hypatia* 27, no. 3: 567–82.

"Status of Spring." n.d. USA National Phenology Network. https://www.usanpn.org/news/spring.

Stephens, Joshua. 2019. "Okjökull Remembered." NASA Visible Earth, August 9, 2019. https://visibleearth.nasa.gov/images/145439/okjokull-remembered.

Stewart, George R. 2008. *Names on the Land: A Historical Account of Place-Naming in the United States*. New York: New York Review of Books.

Stoller, Leo. 1956. "A Note on Thoreau's Place in the History of Phenology." *Isis* 47, no. 2: 172–81.

"Survival by Degrees: 389 Bird Species on the Brink." 2019. Audubon Society. https://www.audubon.org/climate/survivalbydegrees.

Thoreau, Henry David. 1854. *Walden; Or, Life in the Woods*. Boston: Ticknor and Fields.

Thornton, Thomas F. 2008. *Being and Place among the Tlingit*. Seattle: University of Washington Press.

Todd, Zoe. 2015. "Indigenizing the Anthropocene." In *Art in the Anthropocene: Encounters among Aesthetics, Politics, Environments and Epistemologies*, edited by Heather Davis and Etienne Turpin, 241–54. London: Open Humanities Press.

Tonino, Leath, and Michael Soulé. 2018. "We Only Protect What We Love: Michael Soulé on the Vanishing Wilderness." *The Sun Magazine*, April. https://www.thesunmagazine.org/issues/508/we-only-protect-what-we-love.

Tronto, Joan. 1993. *Moral Boundaries: A Political Argument for an Ethic of Care*. New York: Routledge.

Van Dooren, Thom. 2014. *Flight Ways: Life and Loss at the Edge of Extinction*. New York: Columbia University Press.

VanHaitsma, Pamela. 2019. "Digital LGBTQ Archives as Sites of Public Memory and Pedagogy." *Rhetoric & Public Affairs* 22, no. 2: 253–80.

Vatz, Richard. 1973. "The Myth of the Rhetorical Situation." *Philosophy & Rhetoric* 6, no. 3: 154–61.

Vick, R. Alfred. 2011. "Cherokee Adaptation to the Landscape of the West and Overcoming the Loss of Culturally Significant Plants." *American Indian Quarterly* 35, no. 3: 394–417.

"Volunteer Scientists." n.d. USA National Phenology Network. https://www.usanpn.org/partner/volunteer-scientists.

Wallace-Wells, David. 2019. *The Uninhabitable Earth: Life After Warming*. New York: Tim Duggan.

Watson, Janell, and Jane Bennett. 2013. "Eco-Sensibilities: An Interview with Jane Bennett." *Minnesota Review* 81:147–58.

Weber, Elke, Irena Bauman, and Olafur Eliasson. 2014. "Can Art Inspire Climate Change Action? An Ice Installation Aims to Do Just That." *Guardian*, October 23, 2014. https://www.theguardian.com/.

Webster, Robert M., and Bruce Erickson. 1996. "The Last Word?" *Nature* 380: 386.

Wells, Justine, Bridie McGreavy, Samantha Senda-Cook, and George F. McHendry Jr. 2018. "Introduction: Rhetoric's Ecologies." In *Tracing Rhetoric and Material Life: Ecological Approaches*, edited by Bridie McGreavy, Justine Wells, George F. McHendry Jr., and

Samantha Senda-Cook, 1–36. London: Palgrave Macmillan.

Whale, Helen, and Franklin Ginn. 2017. "In the Absence of Sparrows." In *Mourning Nature: Hope at the Heart of Ecological Loss and Grief*, edited by Ashlee Cunsolo and Karen Landman, 92–116. Montreal: McGill–Queen's University Press.

White, Lynn, Jr. 1967. "The Historical Roots of Our Ecological Crisis." *Science* 155, no. 3767: 1203–7.

Whitmore-Williams, Susan Clayton, Christie Manning, Kirra Krygsman, and Meighen Speiser. 2017. "Mental Health and Our Changing Climate: Impacts, Implications, and Guidance." Washington, DC: American Psychological Association, Climate for Health, and ecoAmerica. https://www.apa.org/news/press/releases/2017/03/mental-health-climate.pdf.

"Who We Are, What We Do." 2020. Aldo Leopold Foundation. https://www.aldoleopold.org/about/our-work/.

Whyte, Kyle Powys. 2020. "Against Crisis Epistemology." In *Routledge Handbook of Critical Indigenous Studies*, edited by Brendan Hokowhitu, Aileen Moreton-Robinson, Linda Tuhiwai-Smith, and Steve Larkin, 52–64. London: Routledge.

Wichelns, Herbert A. 1925. "The Literary Criticism of Oratory." In *Studies in Rhetoric and Public Address in Honor of James Albert Winans*, edited by A. M. Drummond, 181–216. New York: Century.

Williams, Joy. 2006. "Fosse." In *Home Ground: A Guide to the American Landscape*, edited by Barry Lopez and Debra Gwartney, 189. San Antonio: Trinity University Press.

Wilson, Edward O. 1984. *Biophilia: The Human Bond with Other Species*. Cambridge, MA: Harvard University Press.

Windle, Phyllis. 1992. "The Ecology of Grief." *BioScience* 42, no. 5: 363–66.

"Working Group on the 'Anthropocene.'" n.d. Subcommission on Quaternary Stratigraphy. http://quaternary.stratigraphy.org/working-groups/anthropocene/.

Wright, Sarah D., and Nina Leopold Bradley. 2008. "Thinking Like a Flower: Phenology and Climate Change at the Leopold Shack." In *The Vanishing Present: Wisconsin's Changing Lands, Waters, and Wildlife*, edited by Donald M. Waller and Thomas P. Rooney, 41–53. Chicago: University of Chicago Press.

"Wyalusing State Park." n.d. Wisconsin First Nations. https://wisconsinfirstnations.org/wyalusing-state-park/.

Wynia, Rich. 2009. "Plant Fact Sheet for Compassplant (*Silphium laciniatum* L.)." US Department of Agriculture Natural Resources Conservation Service. https://plants.usda.gov/DocumentLibrary/factsheet/pdf/fs_sila3.pdf.

Yancy, George, and Judith Butler. 2015. "What's Wrong with 'All Lives Matter.'" *New York Times*, January 12, 2015. https://opinionator.blogs.nytimes.com/2015/01/12/whats-wrong-with-all-lives-matter/.

Yancy, George, and Judith Butler. 2020. "Judith Butler: Mourning Is a Political Act Amid the Pandemic and Its Disparities." *Truthout* (blog), April 30, 2020. https://truthout.org/articles/judith-butler-mourning-is-a-political-act-amid-the-pandemic-and-its-disparities/.

Yong, Ed. 2019. "The Last of Its Kind." *Atlantic*, July. https://www.theatlantic.com/.

Yusoff, Kathryn. 2011. "Aesthetics of Loss: Biodiversity, Banal Violence, and Biotic Subjects." *Transactions of the Institute of British Geographers* 37, no. 4: 578–92.

Index

A

B

C

H

M

N